TRACTORS
OF THE WORLD

TRACTORS
OF THE WORLD
A FULLY ILLUSTRATED HISTORY

MIRCO DE CET

Arcturus

Although the majority of pictures come from the author's archives, he and the Publisher would like to thank the following people and organizations for their kind contribution:

AGCO – Massey Ferguson, Challenger, Fendt, Valtra
John Deere – ASM Public Relations, UK
SAME Deutz-Fahr
CNH – Case IH, New Holland
ARGO – Landini, Mc Cormick
CLAAS
JCB
Kubota
US Library of Congress, Washington DC, USA
Wisconsin Historical Society
University of Guelph, Ontario, Canada
Ford UK, Brentwood, England
The Henry Ford, Dearborn, USA
Canal Farm Vintage Centre, Harby, Leicestershire, England
HJR Agri Oswestry Ltd, Oswestry, England
Andrew Morland Archive: P25

ARCTURUS

This edition published in 2015 by Arcturus Publishing Limited
26/27 Bickels Yard, 151–153 Bermondsey Street,
London SE1 3HA

ISBN: 978-1-78404-474-9
AD004324UK

Printed in China

CONTENTS

Farmyard revolution

Tractors changed everything, from the size of farms and how the countryside looks to helping to make possible the mass exodus from rural areas into the cities. Before tractors, work on the land was carried out by draft animals – oxen, mules and horses – and impoverished men and women who toiled long and hard in the fields, planting and growing crops.

At the beginning of the 18th century, agriculture had changed little in Europe since the Middle Ages. Sowing was done by hand, cultivating by hoe, while hay and grain cutting was carried out manually with a sickle. By 1830, using a walking plough, it took about 250–300 man-hours to gather 100 bushels of wheat (one bushel is equivalent to 8 gallons dry volume), a task which might take a modern tractor just one or two hours.

With the arrival of mass-produced reapers, steel ploughs and other implements pulled by animals, productivity increased and farming became a little less back-breaking. By the 1870s, some US farms were using gargantuan steam traction engines – resembling early locomotives that had run off the rails – to power production, but their size (some weighed more than 14,000 tonnes), unsuitability for wet terrain and mechanical failings caused many problems. Not only did they get bogged down in mud, but there was the ever-present fear that they might explode.

Soon the race was on to reduce the size and cost of tractors, to make them cheaper and more efficient than the animals they were set to replace. Back then, horses and mules were expensive assets, consuming more than 20 per cent of the food that they helped farmers to produce.

In 1892 a man by the name of John Froelich built what is generally recognized as the first gasoline farm vehicle (though the word 'tractor' was yet to be invented). Froelich's company, the Waterloo Gasoline Traction Engine Company, came up with several other examples, but the company was sold on

The Fordson stand at a trade fair in 1922 in the USA. Between 1921 and 1927, 25,000 Fordsons were exported to the Soviet Union, and a Fordson-Putilovets plant was built in Leningrad in 1924 as part of the drive towards farm collectivization.

to a man who would become a giant in all things related to agriculture – John Deere.

Another significant player in the tractor business, someone who had originally made his name with automobiles, was Henry Ford. In 1922 he pulled the rug out from under the competition by reducing his tractors' price from $625 to $395, and by 1925 he had tied up half the tractor market of the day, selling some 100,000 Fordsons per year. What he had done with his automobiles, he was now doing with his tractors – mass production on a scale never seen before. These were just two of many pioneers.

The Massey Manufacturing Company and A. Harris, Son & Company merged in 1891 to become Canada's largest agricultural firm, Massey-Harris (later to become Massey Ferguson). Charles Hart and Charles Parr produced their first gasoline engine in 1902 and were credited with introducing the term 'tractor'. The Holt Manufacturing Company – Benjamin Holt was later to become one of the founders of

A Ford 5000, known as the Ford Major in the USA, a mid-range tractor made for mid-sized farms in the 1960s/1970s.

the Caterpillar Company – sold its first steam-powered tracked tractor in 1904. There were, of course, many who were less well known, but who also contributed massively to the progress of farm machinery.

With GPS self-steering systems, 32-speed transmission, power management, six-cylinder turbocharged, low-emission engines and precision planting, the modern tractor is a highly sophisticated machine. This makes it particularly effective on large farms where the computer can work out such things as the most economical way to plough a field, thus making the job as efficient and time-saving as possible. These days ploughing can even be done at night. All this is crucial for farmers, whose survival has always depended on matching what the market place is prepared to pay for their goods. Nowhere is the phrase 'time is money' more relevant.

The once humble tractor has never been as sophisticated, speedy and powerful as it is today. And early tyros such as Benjamin Holt, Daniel Massey and Jerome Increase Case would be astonished to see what advances have been made since they first tinkered with their radical new inventions. No matter how small the advance or how tiny the idea, in their own way each one contributed the inspiration and genius to make the tractor what it is now: a universal workhorse that enables food and goods to be delivered in sufficient quantity to satisfy the demands of an ever-growing market.

A farmer prepares the soil in his field using a John Deere with a sophisticated set of attachments.

Chapter 1

THE AGE OF MECHANIZATION

The ideal tractor is adaptable, speedy, cheap to run and resilient, exactly the opposite of the lumbering, clanking monsters, belching smoke and cinders, that first took on farm work. Steam-driven traction engines needed a constant supply of coal and water, and they were often forced to make huge detours because their massive weight would crush local bridges. Nonetheless, on large farms they often proved more economical than horses and mules.

Up to the late 1800s, agriculture depended on the physical strength of the people that worked the land. It was a labour-intensive environment and those who tilled the fields worked long hours for little reward.

Men and women – and often children too – would start work at daybreak and only finish once the sun had gone down. Tools were operated manually and larger implements were pulled by animals, mostly horses or bulls.

Looking after the animals that supplied the power to pull the implements was an added problem. They needed feeding, watering and cleaning all year round. So it was a full-time occupation that needed dedication and willpower.

Some of this changed when the steam traction engine was introduced, although these huge contraptions also

had their problems: they needed time to build up a head of steam before even starting work, and then at the end of the day they needed to 'steam down', be thoroughly cleaned and carefully maintained.

Many farmers couldn't afford one, or perhaps they had very soft soil where the traction engine had difficulty coping. These machines were extremely heavy and travelled very slowly and many were simply used to power another machine, such as a thresher or woodcutter.

The ploughing steam traction engine was a particular type of traction engine which was distinguished by a large-diameter winding drum, driven by separate gearing from the steam engine, and usually positioned under the boiler. Fixed to the drum was a length of cable that wound in and out.

This could be attached to a plough, which would shuttle back and forth between a pair of ploughing engines, hauled across the field between the two. This method was fine for smaller fields, but could not cope with the huge farms common in the USA, where the distances were much greater.

Despite all its drawbacks, there is no doubt that when the steam-driven traction engine was introduced, it did make quite an impact on farm life and, like every new invention, it too developed and changed as time went on.

American pastoral

ABOVE: An engraving from 1867, entitled 'Harvest on historic fields – a scene at the South', features a southern farm in North America during the Civil War, possibly in the wake of the Shenandoah Valley Campaign of 1864.

Three years ago the battle's breath
Swept fiery-hot across the plain;
And steadily the reaper Death,
With cruel carnage in his train,
Marched through the serried ranks, that stood
Unwavering, and cut them down;
While field and farm and hill and wood
Grew dark beneath the battle's frown.

To-day another harvest stands
Where once Death trod the bleeding plain,
Ripe for the reapers' ready hands
That bind in sheaves the golden grain.
Afar the sheltered farm-house sleeps,
Embowered in shade; while o'er the mound,
With pitying growth, the wild-vine creeps,
Where rifles rang with deadly sound.

ABOVE: A print from 1854 shows farmers and fieldhands cutting, bundling, and loading sheaths of wheat on to a wagon, a typical scene of the period. The days were long and it was punishingly hard work – a far cry from the methods of today.

LEFT: Two verses from a poem written at the time to commemorate the restoration of peace and order to the countryside, once farmers had returned to their fields from the front line of battle.

Yoked to the plough

It is generally understood that ancient Egyptian civilization was so successful because they farmed the fertile soil around the Nile, producing their own food, as depicted here.

Ploughs have been around for so long it's difficult to say when they were first used; they certainly go back to the days of the Ancient Egyptians. The plough in those days was a simple implement and didn't advance much until the 1800s.

Many early ploughs were merely long pieces of shaped wood with a badly fitting iron point on the end. Generally it was local blacksmiths who made ploughs, but to their own design and without much quality control. Hauled by oxen or horses, the ploughs could cope with soft ground, but heavy, wet conditions caused serious problems.

In 1797, Charles Newbold, of New Jersey, received the first patent for a one-piece, cast-iron plough, which he had designed. Sadly he was unable to sell it due to local superstition that the iron plough would poison the ground. David Peacock was subsequently issued with a patent in 1807 for a three-piece iron plough. Newbold then sued Peacock for patent infringement and won $1,500.

Jethro Wood, a blacksmith from Scipio, New York, received a patent for his three-part, cast-iron plough in 1814 and another in 1819. By now, farmers were less sceptical because they could see the advantages of these devices; for instance, broken parts could be changed without having to buy a whole new plough.

Around this time, William Parlin moved to Canton, Illinois where he began his innovative work as an apprentice for Robert Culton. They produced their first steel plough in 1842. These early ploughs, which had wooden mouldboards with metal plates, proved a great help to local farmers.

William Parlin and brother-in-law William J Orendorff joined forces in the 1850s to form their company P&O. They manufactured a variety of ploughs like this one from a catalogue of the period.

In 1852, Parlin's brother-in-law William Orendorff became his partner and in 1860 they renamed the company Parlin and Orendorff (P&O). They produced many successful farm implements and also invented such machines as the walking cultivator in 1856, the shovel plough in 1857, the riding cultivator and the first lister (a double mouldboard plough) in 1865. Altogether they produced more than 1,400 different types of farm implement, and none was more popular than the Canton Clipper Plow which sold in great numbers.

After Parlin and Orendorff died, their two sons took over the management and by 1919 the factory ranked number one in the plough manufacturing industry. It was eventually sold to the International Harvester Company, and the Canton Works, as it was known, kept producing farm implements until 1983, when it sadly shut down.

In 1833, John Lane arrived in Homer Township, Illinois, and soon began to attack the problem of building a steel plough. It was difficult to work the steel into a curved shape because of its tendency to crack, but Lane solved this problem by cutting a saw blade into strips, and then welding the strips to a thin piece of softer iron. The plough could then be hammered into the desired shape, then ground and polished until it was ready to do the job.

James Oliver, a Scottish immigrant, bought a foundry in South Bend, Indiana, and in 1855 began experimenting with improved farm plough designs. The Oliver Chilled Plough, for which Oliver registered 45 patents during his lifetime, would help his company to become one of the largest in Indiana and one of the world's biggest producers of farm ploughs during the late 19th century.

In 1837, John Deere developed and marketed the world's first self-polishing, cast steel plough. It was not long before two or more ploughs were combined into a single implement, allowing more work to be carried out by the same person. When the 'sulky plough' came along, it allowed the ploughman to ride rather than walk, and by the end of the 19th century giant steam traction engines were pulling ploughs, which in turn brought new problems to be solved.

An early advertisement from the Oliver Chilled Plow Works. With the success of his plough, the Oliver Company moved into other areas of farm equipment, including tractors.

Massey meets Harris

The Massey family emigrated from England to America in 1630, and 1798 saw the birth of Daniel Massey, a name that endures today in the farming world. The Masseys moved to Canada and Daniel set up as a blacksmith in Newcastle, Ontario where he began making agricultural implements in 1847.

In 1851, Daniel's son Hart joined the company, then operating under the name of the Newcastle Foundry and Machine Manufactory, CW (Canada West). Soon they were producing and licensing all kinds of agricultural products. After a fire destroyed their warehouse, the company took on the name of the Massey Manufacturing Company, with Hart as president.

With the success of the open-end binder in 1890, Hart Massey was prompted to propose a merger with their rival Alanson Harris. It was this momentous move that culminated in what was to become, and remains to this day, one of the giants of the farming world. Massey-Harris (later to become Massey Ferguson) was created in 1891.

An early advertising brochure from the Massey Manufacturing Company.

Seen here is the *Massey Illustrated Journal* of 1890, advertising the Toronto Mower.

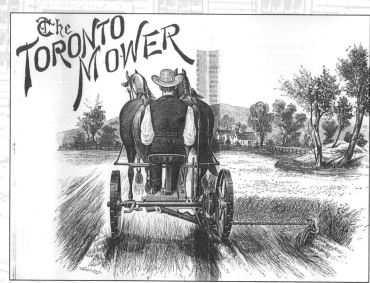

Massey-Harris

Alanson Harris was born in 1816 and became a farmer and saw-mill owner. He bought a foundry in Beamsville, Canada in 1857 and started manufacturing farm implements. He was joined by his son John in 1863, who became a partner in the business, A. Harris, Son & Co.

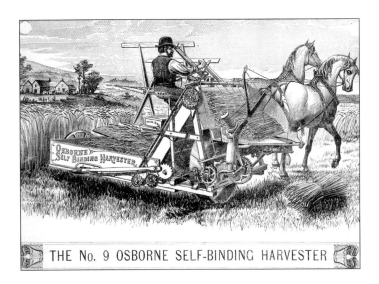

THE No. 9 OSBORNE SELF-BINDING HARVESTER

ABOVE: Harris products were snapped up as far away as Santiago in Chile.

TOP LEFT: It is difficult in today's world of hi-tech machinery to understand just what a leap forward in farming the simple self-binder was.

LEFT: The Kirby two-wheel mower was a great success. It looks very relaxed, but the operator was constantly having to control the horses and make sure he was cutting in the correct way.

The company flourished by acquiring Canadian rights to American products, such as the Kirby mower and reaper. It also developed its own products and in 1872 moved to Brantford, where the company started marketing its products throughout western Canada. The Brantford binder became one of its best-selling products and in 1890 Harris introduced the open-end binder, which was a modified version of the Osborne design. Competition was stiff and their biggest rival was the Massey Manufacturing Company. When Hart Massey suggested a merger, Alanson – who had recently lost his son – decided that it would be for the best, and so the two companies became one. Thus the Massey-Harris Company Ltd came into being.

Traction engines

Methods of ploughing varied according to the fields that were being worked. In England, for example, where farms and fields were considerably smaller than in the vast open spaces of America, traction engines would often operate in pairs, straddling the field and winding the plough between them. The plough shuttled from one side to the other, pulled along metal cables. As the plough ended one row, the traction engine would shift on to the next, and so on until the farmer was finished. The plough was fitted with a tilting system, so that it could work either way – left to right, or right to left.

On this page you can see a ploughing traction steam engine, which is different to the normal traction steam engine. Generally the boilers were longer and they had the cable winding mechanism under the boiler, as shown here. American ploughing engines didn't necessarily have cables like these because their fields were so vast. They had a long cable that stretched out from the rear of the engine, which would be hitched up to the implement that was being used.

ABOVE: The steel rope would extend from this wheel under the boiler of the traction engine. It could be wound in and out as required by the movement of the plough going back and forth across the field.

RIGHT: A ploughing traction engine, seen with its steel rope tucked under the boiler.

Pulling power

Steam traction engines were used for a variety of work, in particular for agricultural purposes, but also for such things as woodcutting, as seen in this reconstruction of a time gone by. They were generally used as the motive power, allowing a separate piece of machinery – a thresher or woodcutting machine, for example – to work via a belt that would be connected between the traction engine and whatever machinery was being used. This method continued even when tractors took over from traction engines; they too had the ability to work other machinery via a belt.

The wheel on the side of the traction engine drives the belt, which in turn is attached to the woodworking machinery. When the wheel on the traction engine turns, it winds the belt and thus turns the woodcutting machine. These methods were adapted for many other farming activities, such as working threshing machines.

J I Case

Jerome Increase Case was born to a Williamstown, New York farming family. He built his first portable steam engine in 1869, which was used to power wheat threshers. As can be seen on the poster advertising J I Case products of the period, he won first place at the 1878 Paris Exposition in France for his thresher, which would be the first of thousands that would later be exported internationally. In 1865, Case introduced the eagle company logo, which was based on Old Abe, a Wisconsin Civil War Regiment's mascot.

ABOVE: An early advertisement for the J I Case company, featuring the range and scope of its agricultural machinery.

RIGHT: Old Abe, as he was known, became the J I Case emblem.

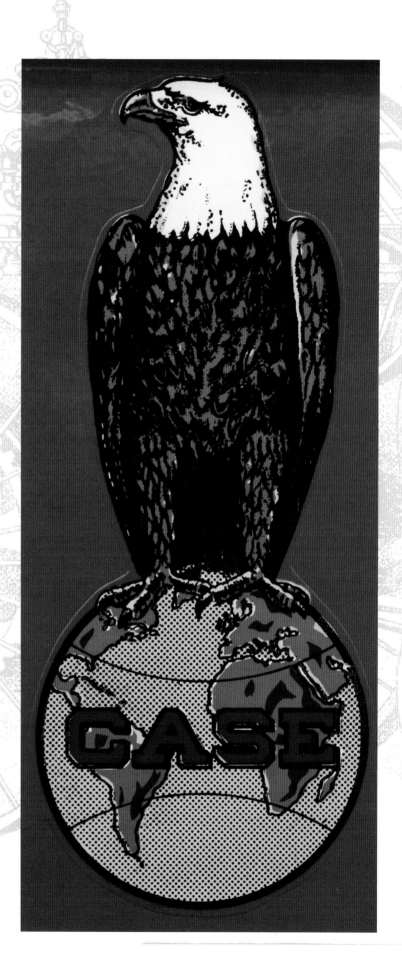

The Case ploughing traction engine shown on this page was made in 1902 and not only carried out ploughing duties but also did a range of other work, including threshing and woodcutting. It is generally known as a direct ploughing engine, meaning that the implement was dragged behind it, rather than via a wire cable tucked under the boiler and working in tandem with another traction engine. This other method, used extensively on the smaller farms of England, is generally known as indirect ploughing.

RIGHT: A driver's point of view – these machines may have moved slowly, but there was a great deal of skill in driving them.

BELOW: A beautifully restored Case traction engine. This one is a ploughing machine designed for the larger fields of the USA, and that's why it is not equipped with the steel rope under the boiler.

A day's threshing

An early agricultural photograph taken between 1910 and 1919, probably in North Dakota. On the left, below, is a water wagon, which sits next to a Case steam engine, feeding it water through a hose. The belts from the steam traction engine extend to the threshing machine and power it in the production of straw. On the right of the steam engine is a full haywagon and further to the right are two partially filled haywagons side by side. The horses that pull them wait patiently to play their part in the process. To the right of the threshing machine is a grain wagon. A man, possibly the farmer, sits in his Model T Ford truck just to the right of the steam engine, apparently keeping an eye on proceedings. It all looks so tranquil but in reality this was very hard work and the days were long, dirty and exhausting.

Chapter 2

THE FIRST TRACTORS

The Charter Gasoline Engine Company of Sterling, Illinois has been credited by experts with being the first company to use petrol as fuel. It then adapted its engine to fit on a Rumely traction engine chassis, and by 1889 had produced six of the machines, which made it a true pioneer in the field of working, petrol-powered traction engines.

And so with progress like this, by the early part of the 1900s farmers were starting to have a real alternative to the heavy and rather clumsy steam engines.

The internal combustion engine, which could be powered by either kerosene (paraffin) or petrol, started to make its mark around 1905. These engines were much less time-wasting – there was no need to 'steam up' first thing and 'steam down' at the end of the day – and the fuel was far less bulky than the mounds of coal, wood or straw that fuelled the fires of the steam

engines, and so everything suddenly became smaller and more manageable.

Although at first the petrol engine was large and far from reliable, by 1912 there were more petrol vehicles available than steam, and by the end of the First World War (1918), petrol had pretty much replaced steam for general mechanized work.

It has to be noted, though, that all this new technology was not available to every farmer, only to the lucky few who could afford these new vehicles. The average farmer struggled on with his horses and other animals for quite some time. It wasn't until 1941 that the majority of prairie farmers in the USA owned tractors, for example. But there is no question that petrol power, and the new vehicles that used it, made a huge difference to the farmer and the way he worked his land.

John Froelich

John Froelich used to spend his autumns taking his steam engine, thresher and a team of helpers around the farms of Langford, South Dakota to work the fields.

Even though the steam traction engine was better than doing things manually, it was still very slow, heavy, bulky and difficult to manoeuvre. The sparks and heat from the engine were a constant danger, as they could easily set fire to the grain. If that happened, and there was even the slightest breeze, disaster was inevitable.

Froelich started looking into other ways of powering his machine and decided a petrol engine would make all the difference. In 1892, he and his blacksmith Will Mann fitted a vertical, single-cylinder gasoline engine to the running gear of a hybrid steam traction engine of their own design. After hours of adjustments and making special parts, they came up with the Froelich tractor (although the word tractor was not as yet in use).

That year he set off as usual for the fields of South Dakota. After 52 days, he and his team had threshed 72,000 bushels – far more grain in a shorter time than had ever been threshed before. The new petrol-powered machine was a success!

John Froelich's 'tractor', powered by a gasoline engine, an idea he got after successfully mounting such an engine on his well-drilling contraption. He is also credited with having invented a washing machine, dishwasher and dryer and a mechanical corn picker.

The Ivel Agricultural Motor

By November 1901 British inventor Dan Albone had built his prototype for a lightweight petrol-powered, general purpose agricultural vehicle and he filed for a patent on 15 February 1902. Ivel Agricultural Motors Limited was formed on 12 December 1902 and he called his machine the Ivel Agricultural Motor.

The machine was a powerful and compact vehicle, which had a single small front wheel and two large rear wheels. The original engine made by Payne & Co of Coventry, England, was a two-cylinder, four-stroke unit of 2,900 cc (177 ci) and could travel at 5 mph (8 kph). A French-made Aster engine was used after 1906. There was also a pulley wheel which could be used to power a variety of external agricultural machinery via a belt. Many of the 500 machines built were exported around the world.

ABOVE: A great example of an existing Ivel Agricultural Motor, the tricycle-style machine, developed by Dan Albone. The huge grey water tank that helps to cool the engine dominates the vehicle.

RIGHT: Classed back then as a lightweight tractor, the Ivel weighed in the region of 1,650 kg (3,638 lb) and cost about £300 ($450/€330) new. The one shown here was manufactured in 1903 and has been lovingly restored.

Ford's experimental tractor

Henry Ford experimented extensively with tractor designs and produced his first petrol-powered version in 1907, under the direction of chief engineer Joseph Galamb. Never put into production, it was referred to as an 'automobile plow', and from the picture here it is obvious why. It used an engine taken from a 1905 Model B automobile, while the chassis and other parts came from a Ford Model K. Ford was unable to use his own name for his tractors, which were soon rechristened the Fordsons (Ford & Sons). This name was first adopted in February 1918.

Henry Ford sits proudly on his 'automobile plow', just one of many experimental machines he designed. His stated aim was 'to lift the burden of farming from flesh and blood and place it on steel and motors', which he did with great success.

The Holt tractor

Around 1883, Benjamin Holt invented his first horse-drawn Link-Belt Combined Harvester in Stockton, California. In 1890, he produced a steam traction engine, 'Old Betsy', which could harvest fields for one-sixth of the price of a horse-drawn combine.

In 1909, a company called R Hornsby & Sons, in Grantham, England demonstrated the 'caterpillar track' to the British Army. The idea was admired but not taken up, so Hornsby sold the patent to Holt, who adapted it for his new generation of tractors. They sold like hot cakes across the US grain belt.

After World War I broke out, the British and French governments put in a large order for Holt tractors. Watching them haul heavy artillery in 1914, Colonel Ernest Swinton had the idea that was to win the war: if you attached caterpillar tracks to an armoured vehicle, you would have a formidable fighting machine. The British tank was born.

The Holt 75 was manufactured between 1913 and 1924. Its track system made it the perfect testbed for tank experiments during World War I.

SPEC BOX

Make/Model	Holt 75
Year	1913–1924
Engine	4-cylinder, gasoline, 1,400 ci (22.9 litres)
Horsepower	75
Power	Drawbar – 50 hp Belt – 75 hp

The Caterpillar Company

When World War I came to an end, Holt Tractors found itself in difficulties since its stock consisted of far too many heavy-duty vehicles unsuited for agriculture. Meanwhile, CL Best had become dominant in the domestic American market. The top two companies in building tracked vehicles were deadly rivals and had spent a combined $1.5 million in lawsuits against each other – $35 million in today's money – since 1905.

In 1919, the CL Best Tractor Company introduced the CL Best Tracklayer, later known as the Caterpillar Sixty, the biggest threat to the Holt 10 Ton tractor. In 1925, the bitter foes had to bury the hatchet as mutual interest forced Best to merge with Holt Manufacturing to become the Caterpillar Company. At this point, the Sixty was renamed and it rapidly became the star performer in the Caterpillar product range.

ABOVE: This is a Caterpillar Sixty with a difference. Before deciding to design their own, Caterpillar experimented with a number of different diesel engines. This model was built in 1928 and used an Atlas Imperial by Henry J. Kaiser, one of only three ever made.

BELOW: The Caterpillar model Sixty was based on the Best Sixty, which had been in production since 1921. The diesel version made its debut in 1931.

SPEC BOX	
Make/Model	Caterpillar Sixty
Year	1925–1931
Engine	Caterpillar 4-cylinder gasoline, 1,128.2 ci (18.5 l)
Horsepower	73
Power	Drawbar – 40 hp Belt – 73 hp (tested)

The Harvesters War

Cyrus Hall McCormick was an inventor and the founder of the McCormick Harvesting Machine Company, and although he is often seen as the inventor of the mechanical reaper, much work was also done by others, including his family members. One of the first public demonstrations of mechanical reaping took place at the village of Steeles Tavern, Virginia in 1831. McCormick was granted a patent for the reaper on 21 June 1834.

CYRUS McCORMICK'S REAPER

LEFT: This print shows a serious-looking Cyrus McCormick holding a paper called the *Prairie Farmer*.

BELOW: A farmer is depicted sitting happily on his McCormick harvester with twine binder. This was the first binder that was able to tie the bundles with twine rather than it having to be done manually.

During the mid-1800s an agricultural war took place in the USA, which was known as 'the Harvesters War'. This involved a number of farm equipment manufacturers who were building harvesting machines like their lives depended on it, each company seeking to outdo the others. Sadly production rapidly outstripped demand and inevitably the price of binders plummeted.

Something had to give and so the banking company J P Morgan brokered a merger of the five rival companies in 1902: the McCormick, Deering and Milwaukee Harvester companies, Plano Manufacturing Co and Warder, Bushnell & Glessner (Champion Harvesters). The result was the creation of the mighty International Harvester Company.

LEFT: Advertisements from two of the biggest players in the 'Harvesters War', McCormick Harvesting Machine Co and Wm. Deering & Co, who were eventually brought together in a mass merger with three other rivals to form the International Harvester Company.

BELOW: Named the Mogul 12-25 in 1912, this vehicle actually saw the light of day back in 1911. It would become the first vehicle from International Harvester to be produced as a lightweight design, and continued to be manufactured up to 1918.

International Harvester

International Harvester produced a range of large petrol-powered farm tractors, under the Mogul and Titan names, which were sold through the McCormick dealerships. They were mainly used for pulling ploughs and for belt work on such machines as threshers and woodcutters. These machines enjoyed varying success, but by the mid-1910s farmers were looking towards much lighter and cheaper models, which were starting to appear on the market.

By early 1913, even the smaller farmer was keen to buy a tractor, but the large unwieldy machines that were around were either prohibitively expensive or just too cumbersome. So when International Harvester introduced the Mogul 8-16 in 1914, it couldn't help but sell well. It was much smaller than the larger steam engines and with its two small front wheels placed close together, it was considerably more manoeuvrable. This was a true lightweight tractor for the period, utilizing a 661 ci, single-cylinder engine, and having a two-plough rating.

In 1916 International Harvester introduced the Titan 10-20 model, which became very popular and was often the first tractor many farmers were able to afford.

ABOVE: International Harvester really hit the spot when it introduced the 8-16. Its single-cylinder, hopper-cooled engine design was simplicity at its best.

SPEC BOX

Make/Model	Mogul 8-16
Year	1914–1916
Engine	International Harvester, single cylinder, 603ci (9.88 litres)
Horsepower	8–16
Power	Drawbar – 8 hp Belt – 16 hp

LEFT: Besides being able to pull implements from behind, the Mogul 8-16 had a belt pulley by the side of the engine, clearly visible here, which could be attached to a belt to motivate other external machinery.

ABOVE: The operator's seat looks well sprung and accommodating, but after a day in the fields it could get very uncomfortable! The model seen here has rear mudguards; these were introduced from 1919 onwards.

LEFT: The International Harvester 10-20 Titan tractor was unbelievably popular with farmers, and between 1916 and 1922 the company built over 78,000 units. Like its predecessor, it used a two-cylinder engine.

The Waterloo Boy

In 1893, after John Froelich invented the first practical, petrol-powered tractor, he and a group of Iowa businessmen created the Waterloo Gasoline Traction Engine Company, with a view to putting it into production. Sadly, the tractor was not a great success and, of the four made, only two sold – and even they were returned by their unhappy owners. In 1895 John W. Miller bought the company, renamed it the Waterloo Gasoline Engine Company, ceased making tractors and concentrated on making stationary gasoline engines.

It was not until 1911 that the company once again decided to start designing and making tractors. It had great success with the Waterloo Boy model LA tractor in 1913 and the model R, a single-speed machine, introduced in 1914. Production of the model R finished in 1918, but in the meantime a model N, with two-speed gearbox, was introduced to the market in 1916. In 1918, Deere & Company bought the Waterloo Gasoline Engine Co and the new outfit was known as the John Deere Tractor Company.

ABOVE: Despite the company name, these tractors ran on kerosene.

BELOW: The model N Waterloo Boy tractor, which succeeded the model R.

SPEC BOX

Make/Model	Waterloo Boy model N
Year	1916–1924
Engine	2-cylinder, 465 ci (7.6 litres)
Horsepower	12
Power	Drawbar – 15.98 hp (tested) Belt – 25.97 hp (tested)

Seeing double

Looking just like the model R Waterloo Boy tractor, this is the Overtime, and yes you could be forgiven for thinking that they were one and the same. In fact, you would be quite right. The Waterloo Boy was so successful that from 1917 it was exported to Denmark, France, Greece, Ireland, South Africa and England. These tractors received new livery and became known as the Overtime tractor. This was a reference to the company that imported the machines – the Overtime Tractor Company, London.

Waterloo Boy model R and model N tractors were shipped to England in dismantled form. At the Overtime Tractor Company in London, they were given a new paint job and assembled before being put on sale under a new name.

The Rumely Prototype

The Advance-Rumely Oil Pull is a favourite with many tractor collectors and enthusiasts. The company has roots that date back to 1852, when Meinrad Rumely opened his blacksmith shop in Portland, Indiana. Here he made threshing machines, which became instantly popular and sold widely.

It was Meinrad's grandson Edward who was the main instigator of the move into tractor construction. He gathered together the right people and knowledge to create the company's first prototype in 1908. The tests for the machine went well and a factory was set up. The company sold its first machine just two years later in 1910.

Rumely went on to acquire many smaller companies and produced a plethora of tractors under the Oil Pull designation. Although the company experienced financial difficulties in 1913 and subsequently went into receivership, it emerged again under the new name of the Advance-Rumely Thresher Company, but was no longer the property of the family. It was the Great Depression of 1929 that saw the company fall into financial trouble once more, finally being subsumed into the Allis-Chalmers Company.

LEFT: Rumely Oil Pull tractors were easily distinguished by the huge cooling stack at the front of the machine. This was required due to the engine running on kerosene (paraffin), which burned at a much higher temperature than other fuels.

OPPOSITE: The demise of the Rumely Company was partly due to the huge size of their tractors. Technology was moving on and although powerful tractors were needed, there were now smaller and more compact units on the market.

SPEC BOX	
Make/Model	Advance-Rumely X 25-40
Year	1928–1930
Engine	Rumely 2-cylinder horizontal, 601.4 ci (9.9 litres)
Horsepower	25–40
Power	Drawbar – 38.7 hp (tested) Belt – 50.3 hp (tested)

The Parrett Tractor Company

Dent and Henry Parrett founded the Parrett Tractor Company in 1913 in Ottowa, Illinois. The first successful production model appeared in 1915 and was designated the 10-20. These tractors featured a cross-mounted engine with a sideways-mounted cooling radiator. They could be easily distinguished from other models by their large front wheels. The 10-20 was redesigned to become the 12-25 Model E and H tractors, and an even more powerful version, the 15-30 Model K, was also produced. The Canadian company Massey-Harris, having made an agreement to have a tractor built by the Parrett Company, transferred the production plant to Ontario in 1918, where the 12-25 became the Massey-Harris No.1 and No.2.

SPEC BOX	
Make/Model	Parrett model E 10-20
Year	1915–1917
Engine	Buda 4-cylinder
Horsepower	10–20
Power	Drawbar – 10 hp Belt – 20 hp

LEFT: The Buda 4-cylinder engine of the Parrett 10-20 model is seen poking out from the enclosed engine compartment.

TOP: The company moved to Chicago from Ottowa, Illinois, USA, in 1915 and began building the 10-20 there.

The Allis-Chalmers 10-18 was the company's first attempt at designing a tractor. Not a great start, it failed to catch on due to its three-wheel configuration. The design was heavily influenced by the very popular Little Bull tractor of the same period. In fact, a tie-up between the two companies had been talked about, but never happened.

Allis-Chalmers

Edward P. Allis & Co, established in Milwaukee in 1861, built its first steam engine just a few years later. Joined by a number of companies in 1901, including Fraser & Chalmers of Chicago, Allis-Chalmers was formed. But rapid expansion can often come at a steep price, and on 8 April 1912 the company went into receivership.

The Allis-Chalmers Manufacturing Co was formed on 16 April 1913, with Otto Falk as its president. Falk was also largely responsible for the introduction of farm equipment. A prototype of its first true tractor, the three-wheeled, two-cylinder, gasoline-engine 10-18 model, was completed by November 1914. Although serial production began the following year, it was never a great success – farmers were looking for more conventional four-wheel designs by now.

Son of Ford

Seen as the first real lightweight tractor in the world, the Fordson model F was manufactured in 1916. As with the Ford Model T automobile, these tractors were mass-produced and sold at a considerably lower price than the competition. Many farmers, who had been unable to afford it before, were now given the opportunity to buy their own tractor for the first time.

A factory was built at Dearborn, Michigan, and an automobile-style production line was set up to manufacture these tractors. They used a 20 hp, 4-cylinder engine with a 3-speed transmission. The beauty of this tractor lay in the construction. It did away with the usual heavy frame and used the engine, transmission and rear axle as the stressed member of the frame. In this way manufacturing costs were lowered, so the final sale price could be kept to a minimum. This tractor changed many farmers' lives forever: no longer did they need a team of horses to work the fields, and wage bills too could be reduced.

Initial production of the Fordson F was at Dearborn in the USA, and further production started in Cork, Ireland, in 1919. This is the 4-cylinder, 20 hp engine used in the tractor. The ignition system was similar to the one used in the Ford Model T automobile of the period.

SPEC BOX	
Make/Model	Fordson F
Year	1917–1928
Engine	Hercules 4-cylinder (to 1920)/ Ford 4-cylinder
Horsepower	20
Power	Drawbar – 12.325 hp (tested) Belt – 22.28 hp (tested)

Although the Fordson F did suffer some early teething problems, it became a huge seller and took 70 per cent of the market share in its early years. Before the recession in 1925, the 500,000th tractor had already rolled off the production line.

The coffin-style tractor

The Case 12-25 was introduced in 1914 and was a first attempt to woo farmers with a much lighter and smaller machine. The previous J I Case 30-60 tractor had been large and was now old technology. Farmers were desperately looking for something newer, smaller and more manoeuvrable. The 12-25 fitted right into this sector. With its all-enclosed bonnet, it had the look of a large automobile rather than a tractor. It was powered by a two-cylinder, opposed engine.

By 1914, farmers were desperate to get their hands on smaller, lighter and more manoeuvrable tractors.
The 12-25 model from Case was the company's first venture into that segment of the business. It remained in production up to 1918.

ABOVE: The 10-20 Case model was an oddity, with its three-wheel configuration. The cross-mounted engine was the same as the one used in Case automobiles. Even a reduced price could not sell the tractor in sufficient numbers and it was no match for the Little Bull tractor of the same period.

RIGHT: A farmer seated on his Case 10-20, ploughing the rich prairie soil somewhere in South Dakota.

Steady progress

Introduced in 1915, the Case 10-20 horsepower tractor was an unusual design powered by a 4-cylinder, transverse-mounted engine. The steering wheel was mounted on the far right with the driver's seat directly behind it. Thus the driver was positioned behind the right rear wheel, which had a large mudguard for his protection. The left rear wheel could free-wheel, but could also be engaged for extra traction. The one forward gear gave a speed of 2 mph and, although work went at a slow pace, it was a lot less laborious and a bit more comfortable than doing it manually. The 10-20 tractors remained on the Case inventory until 1918, although there were several still available even after that.

SPEC BOX	
Make/Model	Case 10-20
Year	1915–1918
Engine	Case 4-cylinder, transversely mounted
Horsepower	10–20
Power	Drawbar – 15 hp aprox. Belt – N/A

Aultman-Taylor

Aultman-Taylor introduced its first tractor around 1910–1911. The 30-60 model was originally introduced with a square radiator but was changed to the rounded tubular version, as seen in the picture, in 1913–1914. This model was very popular and remained on the books until the company was taken over by the Advance-Rumely Thresher Company in 1924.

A model 25-50 was introduced in early 1915, which was very similar to the 30-60 except that it was smaller and more compact. Both these models had 4-cylinder engines. Other models followed but the company found the going difficult after the Great War and had to be saved by a takeover.

Dual fans were used to drag the air through the 196 2-inch tubes in the huge, round 120-gallon radiator of the 30-60 tractor. Although a small version of the tractor was introduced, Aultman-Taylor machines started to look dated as farmers sought out more compact, efficient machines.

Tractors and tanks

The military tank had not been developed as an all-terrain fighting machine prior to World War I, although a form of armoured fighting vehicle was available to several countries before fighting started. These 'armoured cars' were not practical when it came to trench warfare, or when crossing the type of terrain that was to be encountered during the war.

These early prototypes took their inspiration from the machinery that was to be found on farms. Tractors were the basic templates of design, in particular for the many vehicles that were being developed to run on tracks. The Holt traction engine was a very popular model and several countries used it for pulling field guns and the like.

Smaller in size were vehicles like the one seen on the left. This is a 5-ton US Army Ordnance artillery tractor, photographed in 1918.

By 1919 the desire for larger and more powerful tanks had grown, but although the tank was really starting to make its mark on the field of battle, it still needed to develop a personality of its own.

ABOVE: This recruitment poster shows how quickly the tank had developed from the simple tractor.

BELOW: A US Army artillery tractor of World War I being put through its paces.

Hart-Parr

Charles W. Hart and Charles H. Parr began experimenting with petrol engines in the late 1800s while at the University of Wisconsin at Madison. They formed the Hart-Parr Gasoline Engine Company of Madison in 1897 and later moved their company to Charles City, Iowa in 1901, where they found financing to make their gas traction engines. Here they built the first factory in the USA dedicated to the production of gas traction engines. Their first tractor – they are also credited with coining the name 'tractor' – was the aptly named Hart-Parr No.1 and was built in 1902. Shown here is the Hart-Parr 30, made between 1918 and 1922.

SPEC BOX

Make/Model	Hart-Parr 30	
Year	1918–1922	
Engine	Hart-Parr 2-cylinder 464.6 ci (7.6 litre)	
Horsepower	30	
Power	Drawbar – 19 hp (tested) Belt – 31 hp (tested)	

Alldays & Onions

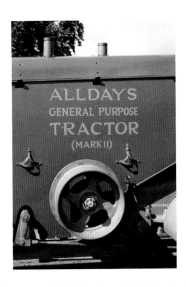

Alldays & Onions are probably better known for their automobiles. Their tractors, of which only one model was ever made, has become a rare collector's piece. One of only two in the UK is shown here – an updated version of the General Purpose Tractor and designated Mk II.

The Alldays & Onions Company came about with the merger of the two companies who went by these names. The new company was based in Birmingham, England and produced a range of products, including turbines, fans, pneumatic hammers and furnaces, as well as a vast range of bicycles, motorcycles and cars.

When it turned its hand to making tractors, there was only one model: the Alldays General Purpose Tractor. In some respects the design was quite forward-looking, with front- and rear-sprung axles and a fully enclosed engine. This gave the appearance of a large car with steel traction engine-style wheels. The tractor was later updated to MK II status.

Chapter 3

BOOM AND BUST

World War II was to signal major changes in the development of the tractor. The First World War had been followed by a period of prosperity and tractor manufacturers took full advantage. Many new models had come on to the market and there was a greater variety than ever before.

Although the smaller companies could not compete with the established companies, they still felt there was room in the market place for their machines. The tractor had by now become smaller, faster and more manoeuvrable, and was thus able to carry out many more tasks. A giant leap forward was achieved with the development of PTO (power take-off), allowing direct transmission from the engine to the farm implement behind the tractor. This also meant that tools being towed could be smaller and lighter too.

It was the Wall Street Crash of 1929 that would determine the level of national economies for the next few years. It signalled the beginning of the Great Depression, which lasted 12 years and affected all Western industrialized countries. In the USA it lasted right up until mobilization for World War II which came at the end of 1941.

During this period, the larger tractor companies consolidated, but many of the small outfits went to the wall. It was a time of tightening belts and making do. Then, as the 1930s were coming to an end, war clouds started to gather once more, and tractor companies found that they were being requested to supply their products to the military, which meant refitting their production lines for the manufacture of guns, tanks and aircraft.

International lightweight

The International Harvester 8-16 junior model was introduced in 1917 as a lightweight tractor. It followed the slightly bulkier and larger Titan 10-20, which also became extremely popular. The 8-16 model used an advanced overhead-valve, 4-cylinder engine, powered by kerosene. It remained in production until 1922. An interesting design feature was the cooling radiator of the 8-16, which was tucked up behind the engine. This gave it protection from the debris being churned up when working.

RIGHT: An umbrella placed at the rear of the tractor was not a rare sight on farms in the American Midwest. The sun was often so strong during the day that it could make even the strongest farmhand ill from the heat.

BELOW: Looking more like an early car design, the 8-16 became a very popular tractor. It was light (only weighing 3,650 lb [1,655 kg]), manoeuvrable and even reliable.

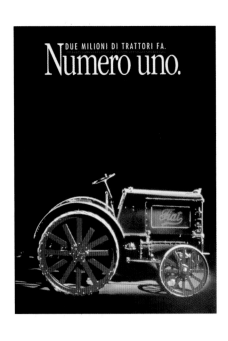

Fiat 702

Fiat Trattori SpA was founded in 1918 as part of the larger Fiat company (Fabbrica Italiana Automobili Torino), which had been founded by Giovanni Agnelli in Turin in 1899. Although the company became well known for its automobiles, it didn't start producing tractors until 19 years into its existence. The company started looking into building tractors as early as 1910, but it was obliged to delay its plans due to the onset of World War I.

The model 702 was the very first Fiat agricultural tractor, as well as the first Italian tractor to be built on an industrial scale. It went into full production at the car and truck plant in Turin, Italy in 1918. It used a 25 hp, 4-cylinder petrol engine, had one reverse and three forward gears and remained in production until 1924.

ABOVE: 'Two million tractors make: Number One,' says the Fiat ad.

RIGHT: An illustration from the Fiat tractor brochure depicting the 1919: Agricultural Tractor Fiat model 702.

BELOW: A model 702 that is still around and still works. These tractors were exported around the world.

The Diesel Horse

Given the nickname Dieselross (Diesel Horse) due to the fact they took the place of the farm animal, Fendt found they had a winner in this little machine, which came fitted with a plough at the rear.

The post-World War I years saw some strange but interesting machines being produced by all kinds of manufacturers. In 1930, Franz Sailer, a farmer and brewery owner from Marktoberdorf, Germany, bought the first Fendt tractor, which was equipped with a diesel engine. He went on to name it 'Dieselross' (Diesel Horse). This was such a great idea that Johann Georg Fendt and his son Hermann took up the name and used it for their line of tractors, thus marking the beginning of the Dieselross series of tractors and, of course, the Fendt name as a major tractor manufacturer. Today, the Fendt brand is a part of the US-based AGCO Corporation, which is the world's third-largest manufacturer and distributor of agricultural equipment.

SPEC BOX	
Make/Model	Fendt Dieselross
Year	1930
Engine	Fendt single cylinder diesel
Horsepower	6
Power	Drawbar – N/A Belt – N/A

The Hürlimann

The Hürlimann company was founded in Switzerland in 1929 by a young farmer, Hans Hürlimann, who had developed a passion for tractors. He made good use of an apprenticeship as a fitter and toolmaker at the machinery works of August Hoegger in Wil (in the Sankt Gallen canton), where in 1926 he became head of sales. He went on to create a company that would win success as a maker of tractors for agricultural, industrial and military purposes, not just in Switzerland but around the world.

Hürlimann produced its first machine in 1926, a single-cylinder gasoline tractor with mower. Soon afterwards in 1930, the 1K10 went on sale. This also used a single-cylinder engine, which produced 10 hp, had three forward gears and one reverse, and could travel at 16 kph (10 mph). As you can see (below), the mower is attached to the side and can be manoeuvred up and down.

Switzerland is not just the place for cuckoo clocks; its tractors were also precision-built, like this Hürlimann 1K10.

The Caterpillar Company

The Caterpillar Company introduced the RD4 in 1936. It weighed around 10,000 lb (4,545 kg) and was powered by Caterpillar's own D4400 engine, an inline, 4-cylinder unit.

It was 1935 when Caterpillar started the naming convention of RD for diesel versions, and R for the regular petrol versions. This would be followed by a number, which would indicate the size of the engine. Just two years later in 1937, the R was removed and just the letter D was used for the diesel versions. The later D4 series engines quickly increased in power, so the number 4 just became a figure of merit rather than indicating actual engine power.

SPEC BOX	
Make/Model	CAT D4
Year	1937–1957
Engine	Caterpillar 4-cylinder diesel
Horsepower	48
Power	Drawbar – 50.18 hp (tested) Belt – 58.88 hp (tested)

Since the late 1930s, the Caterpillar D4 bulldozer has not just been used for industrial purposes but also for military applications. Although these diesel-powered monsters were slow, they were very sturdy and remarkably versatile.

Ferguson-Brown

In 1933, Harry Ferguson successfully demonstrated his prototype Black tractor, as it was known, which he built in Belfast. Once he had overcome the problems it had with the hydraulics, Ferguson agreed with David Brown that they would manufacture the tractor at the Huddersfield works in England, while Ferguson would take on the sales side of the business. And so the Ferguson-Brown Company was born to manufacture and sell the Ferguson-Brown Model A tractor.

The tractor was lightweight and a pretty radical design, but most importantly it was the first to have the three-point hitch, which Ferguson had spent some 30 years developing. This was simply an easy way of attaching the implements to the rear of the tractor, usually requiring only one operator. Implements could also be lifted up, so that if you were ploughing a field, for example, you could lift the plough up at the end of the row without having to rely on someone else to do it.

ABOVE AND OPPOSITE: The basic tractor was fitted with steel spade lug wheels and independent drum brakes. It had three forward gears plus reverse and a top speed of 5 mph. You could upgrade by ordering power take-off, side-mounted belt pulley and pneumatic tyres.

LEFT: The original Ferguson-Brown Model A tractor used a Coventry Climax engine and ran on petrol or paraffin, changed later to a David Brown engine.

The tractor was originally fitted with a Coventry Climax engine but later a David Brown engine replaced it. Sadly the tractor did not sell that well as it was more expensive than the competition, in particular the Fordson. An additional problem was that, clever though it was, any implement had to be made specially to fit the three-point hitch, which just added to the cost.

In 1938, Ferguson took himself off to America to demonstrate his tractor to Henry Ford, who was impressed with the tractor and its three-point hitch. The two made a gentleman's agreement, which was sealed with a handshake. The outcome would be the Ford-Ferguson 9N.

SPEC BOX

Make/Model	Ferguson-Brown
Year	1936–1937
Engine	Coventry Climax E/David Brown, 4-cylinder
Horsepower	20
Power	N/A

Minneapolis-Moline

Three companies were brought together in 1929 to make up Minneapolis-Moline: the Moline Plow Company, the Minneapolis Threshing Machine Company and the Minneapolis Steel & Machinery Company.

Minneapolis-Moline introduced its 'Visionlined' tractors in 1936, the first of which was the Universal 'Z' model. The 'Visionlined' concept was to enable the driver to be more comfortable while also enjoying a better view of everything that was going on around him. The other distinctive element about these tractors was the new Prairie Gold and red colours, which were also introduced in 1936, although not to the 'Z', which remained the Twin City grey until the following year. Famously, there were three recognized shades of Prairie Gold, of which the original was supposedly only used for two years.

The model 'Z' tractors came in standard four-wheel and also a row-crop version. The engine had a detachable head, which gave access to add or remove spacers, which allowed you to adjust the compression ratio. Therefore, if a farmer wanted the extra power gained from petrol fuel, he could adjust his engine to provide that higher compression.

SPEC BOX	
Make/Model	Minneapolis-Moline ZA
Year	1949–1952
Engine	Minneapolis Moline 4-cylinder, 206 ci (3.4 litre)
Horsepower	32
Power	Drawbar – 30.07 hp (tested) Belt – 36.20 hp (tested)

ABOVE: The distinctive Minneapolis-Moline badge.

BELOW: The Z series of Minneapolis-Moline tractors came in various guises. The ZTS was the standard tread version, whilst there was also a ZTU row-crop version, which had the twin front wheels close together.

The Farmall tractor

Farmall was the name given to the first row-crop tractor manufactured by International Harvester. Work on the new tractor started in the early 1920s and it was unveiled in 1924. International was worried by the new front-wheel design – two small wheels placed very close together – and so initially only sold it in Texas, so as to minimize any possible embarrassment should it fail. They need not have worried because the Farmall became extremely popular, as did the new row-crop, front-wheel design.

The original Farmall, retrospectively named 'Regular', was followed in 1932 by the F-20. Other new models followed and these became known as the 'F-series'. These included the Farmall F-30 in 1931, the F-12 in 1932 and the F-14 in 1938.

Built at the Farmall, Rock Island factory in Illinois for less than two years, the Farmall F14 (shown below) varied little from its predecessor, the F12. The tractor was fitted with a 4-cylinder, 113 ci, water-cooled engine, driving through a 3-forward-speed-and-one-reverse-gear transmission.

The Farmall F14 was only manufactured for one year; production started in 1938 and finished in 1939. Various chassis designs were available for this model.

David Brown

During World War II, David Brown concentrated on the manufacture of gears for tanks and other vehicles, but also carried on making tractors for the military and other enterprises. The David Brown VIG 1 tractor, an industrial version of the VAK 1 tractor, was used during the war for towing aircraft and bomb transport trolleys. The 'tug', as most people refer to it, was mainly used by the Royal Air Force, but many were also used for moving large items around warehouses and factories. The one shown here was used by the Marconi factory in the UK, which today is a part of the Swedish cellphone firm Ericsson.

TOP: With its wrap-around bodywork and car-type tyres, the industrial 'tug' was very different from other tractor models.

RIGHT: Shown here is the powerful David Brown 37 hp-engine, which could be run on petrol or paraffin.

SPEC BOX	
Make/Model	David Brown VIG 1 'tug'
Year	1941–949
Engine	David Brown 4-cylinder 2,523 cc
Horsepower	37
Power	Towing pull – 2.5 tons

Case model D

The J I Case Company increased its production capacity by buying the Rock Island Plow Co in 1937, and just two years later introduced the new streamlined D series tractors. These were also identified by their new colour scheme of Flambeau Red. The D came in a full range of models: DC3 tricycle, DC4 wide axle, DO orchard, DV vineyard, DR rice and the DEX (shown), which was part of the lend-lease agreement and extensively used during World War II. The D used an engine designed and built by Case, rated at 26-32 horsepower. It was also equipped with a mechanical implement lift mechanism.

SPEC BOX	
Make/Model	Case model D
Year	1939–1953
Engine	J I Case 4-cylinder petrol 259.5 ci (4.3 litres)
Horsepower	26–32
Power	Drawbar – 30.67 hp (tested) Belt – 35.36 hp (tested)

The Case model D tractors varied little from the previous C series, and many could only distinguish one from the other by the new Flambeau Red paint job and new styling.

Top of the range in the International Harvester line was the mighty W-9, which used the same engine as the T-9 crawler. Originally it had steel wheels, but rubber tyres also became available later.

McCormick 'standard' W-9

In 1939, International Harvester commissioned the renowned industrial designer Raymond Loewy to design a new line of tractors. The sleek look, combined with other new features, created what is known as the Farmall 'letter series' and the McCormick-Deering 'standard series', which included the W-4, W-6 and W-9.

The model W-9 (shown here) was a four-plough tractor and was the largest of the three gasoline-powered tractors. They were manufactured from 1940 through to 1953 and were very successful, so much so that it isn't that unusual to see one still working today.

The W-9 was powered by an International, 4-cylinder, 335 ci engine. It had five forward speeds and one reverse and was manufactured in several different versions: W-9, WD-9, WR-9, WDR-9, I-9 and ID-9, with a grand total of 37,504 models produced.

John Deere model B

The 2-cylinder 'Letter Series' John Deere model B was manufactured from 1935. It was seen as a smaller tractor that would cater for the row-crop farmers, who did not need a large tractor like the current model A. Although the BR (shown on this page) and BO versions were replaced in 1947 by the model M, the row-crop B continued to the end of the run in 1952.

The model B can generally be split into three groups: the original B was 'unstyled'. This meant that it lacked a grille and other sheet-metal covers for the radiator, which included all machines up to the end of 1938. The 'early styled' machines of 1939 had been redesigned by Henry Dreyfuss, who added grills in front of the radiator along with the sheet-metal surround, giving the tractor a much more streamlined look in line with the automobiles of the period. The model B was restyled once more in 1947 and this became known as the 'late style' for obvious reasons. The most obvious changes were the pressed-steel frame and the cushion-style seat, which now housed the battery box. Interestingly, an electric starter also became standard on this version.

ABOVE: The BR featured here is the standard tread version of the row-crop machine. Several other versions were made, such as the BO orchard and BI industrial version.

BELOW: The original John Deere model B of 1935 had a short frame and 2-cylinder engine. Slight changes were made throughout the B's life, but the 2-cylinder engine remained.

SPEC BOX

Make/Model	John Deere BR 1941
Year	1935–1952
Engine	John Deere 2-cylinder 175 ci (2.9 litres)
Horsepower	19
Power	Drawbar – 24.62 hp (tested) Belt – 27.56 hp (tested)

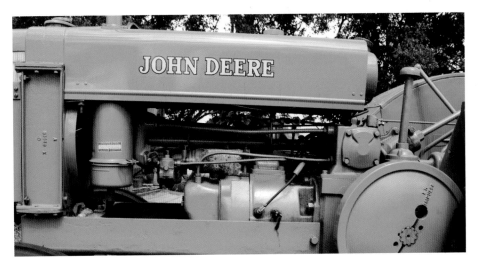

The Oliver crawler

The Oliver Farm Equipment Company was created through the merger of four companies in 1929: the American Seeding Machine Company of Richmond, Indiana; the Oliver Chilled Plow Works of South Bend, Indiana; the Hart-Parr Tractor Company of Charles City, Iowa; and the Nichols and Shepard Company of Battle Creek, Michigan.

It was the acquisition of the Cletrac (Cleveland Tractor Company) that allowed Oliver to add crawlers to its own list of farm vehicles – shown is an Oliver model BD crawler. Oliver, under its own badge, produced both small and large crawlers until 1965, when production was halted. By this time the company had been bought by the White Motor Corporation of Ohio.

ABOVE: A view of the controls from the operator's position. There's no steering wheel here; directional control is via foot pedals and hand levers.

The Oliver BD crawler used either a DJXC or DJXD Hercules 6-cylinder, diesel engine, depending on which year it was manufactured.

David Brown VAC1

Harry Ferguson and David Brown finally had a falling-out and, while Ferguson went off to America to meet with Henry Ford, Brown decided to start making his own tractors. The company had been manufacturing the Ferguson-Brown prior to the disagreement.

And so the first David Brown tractor, which took ideas from the Ferguson-Brown, was named the VAC1 and exhibited at the 1939 Royal Show in Windsor, England. Sadly, production could not start as World War II was imminent and the factory had to prepare for the production of gears for tanks and the like. During this time, it also kept its tractor manufacturing expertise alive by producing aircraft towing and maintenance vehicles.

With these facilities already in place, the manufacture of post-war tractors was made easier. The production line for the VAC1 was organized and the first tractor rolled off in 1945, now designated VAC1A. The engine for the vehicle was a 35-horsepower, 4-cylinder, water-cooled unit, which could work with gasoline or paraffin.

The VAC1 was the first all-David Brown tractor. It came with steel wheels, as here, or with rubber tyres. Giving extra strength with less weight, the frame consisted of engine and gearbox casings, which held other components such as the radiator and also provided the front axle fixing.

The Farmall model A

The Farmall model A, which was one of the new, streamlined 'letter series' (so called because each model was designated by letter) that resulted from International Harvester commissioning Raymond Loewy to give the Farmall range a visual shot in the arm.

International Harvester announced its new small tractor, the Farmall A, in 1939 at a factory preview. It was the smallest tractor the company had ever produced, and was seen as an all-purpose machine for the smaller farm. It had the availability of a variety of implements, thus making farm mechanization a possibility for farmers around the world. Yet this little tractor would also complement the larger machines on any large farm, where smaller duties were needed to be carried out.

The Farmall A had a 4-cylinder, overhead-valve engine with a displacement of 113 ci. The engine also had replaceable cylinder sleeves, precision bearings and a crankshaft that was Tocco-hardened. With its 4-speed transmission, it could reach a speed of 10 mph (16 kph) for highway work and it had a reverse gear. The Farmall model A was only available with rubber tyres and production ended in 1947.

Looking streamlined and elegant, the Farmall M was a big and powerful machine, capable of pulling a three- or four-bottom plough. Over its 14-year production run, more than 290,000 of these machines were built. You could be forgiven for being confused by all the logos and names — McCormick was now a part of the International Harvester Company and Farmall was a brand name. The 'M', of course, denotes the model.

SPEC BOX

Make/Model	Farmall MD
Year	1939–1954
Engine	International Harvester 4-cylinder 247.7 ci (4.1 litres)
Horsepower	34–37
Power	Drawbar – 33.1 hp (tested) Belt – 36.66 hp (tested)

The Farmall model M

The Farmall M was a big and powerful tractor, which became popular with farmers who needed a row-crop machine for their super-sized farms. Production started in 1939 and the M was made in several versions throughout its long and illustrious run. It was finally replaced by the Super M in 1952. The diesel version MD came in 1941, and there was also a high-crop version, the MV. The Australian equivalent, built between 1949 and 1954, was known as the AM. The 1950s saw production cease at the Rock Island Farmall factory in the USA, but in Doncaster, England they continued to manufacture the tractor under the BM designation for a while longer.

The Fordson N

The Fordson model N was a replacement for the model F. Production of the model N started in Cork, Ireland in 1927, but was transferred to Dagenham, England in 1933, and lasted until 1945. The model N used a 26 hp engine, standard rear fenders (mudguards), a higher voltage ignition system and optional pneumatic tyres. These tractors were used extensively during World War II by farmers when food production was vital to the war effort.

SPEC BOX

Make/Model	Fordson N
Year	1929–1945
Engine	Ford 4-cylinder kerosene 267 ci (4.4 litres)
Horsepower	26
Power	Drawbar –18.30 hp (tested) Belt – 29.09 hp (tested)

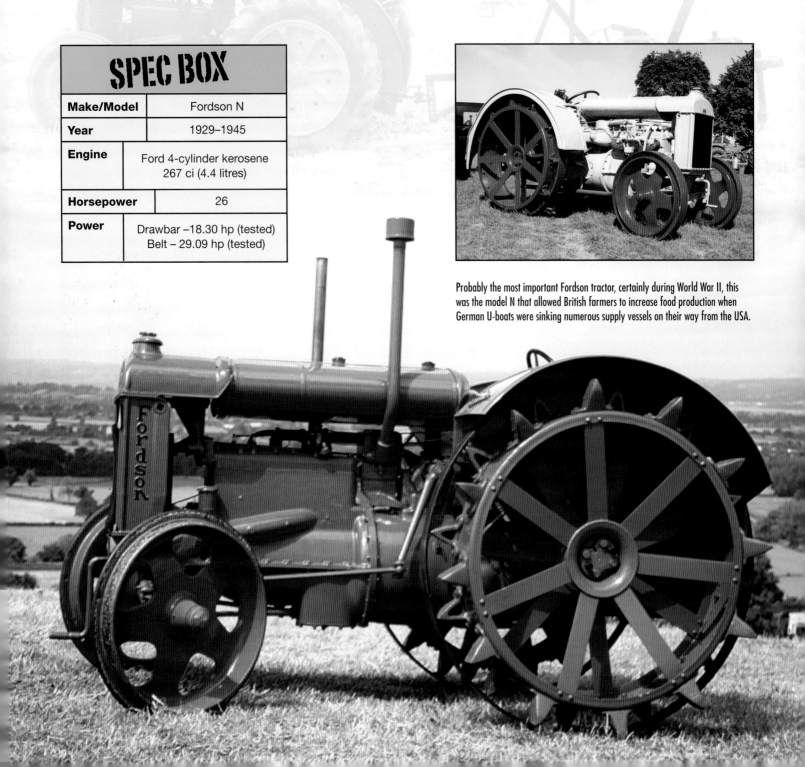

Probably the most important Fordson tractor, certainly during World War II, this was the model N that allowed British farmers to increase food production when German U-boats were sinking numerous supply vessels on their way from the USA.

The Women's Land Army

A typical wartime scene with Land Girls working together to bring in the harvest, while a trusty model N acts as their beast of burden.

The Women's Land Army, or WLA, was a British civilian organization created during World War I and World War II. These women worked in agriculture, replacing the men who had been called up to fight. Commonly known as Land Girls, many of them already lived in the countryside, but others arrived from London and the northern industrial cities.

As the likelihood of war increased, the Government decided it needed to produce more food. To do this, more help was required on the farms, and thus the WLA was created in June 1939. The name was also used in the USA for an organization formerly called the Woman's Land Army of America.

The Case VAC

The new Case VA series, a further development of the V series, was introduced in 1942 and followed on from the R series. It was supplied with many new major components. Initially the tractor used an engine built by Continental Motors, but in 1947 Case took over complete production of the engine in its new engine plant in Rock Island, USA.

The Case VAC, part of the V series of tractors, used many components that were brought in from outside suppliers. For example, the early engines came from Continental Motors of Muskegon, Michigan, while the gearing was brought in from the Clark Company, also in Michigan.

The VA also had several variants of which the VAC was the tricycle row-crop model. The machine pictured here has a 4-cylinder, 125 ci (2,033 cc) petrol engine, which produces 20 hp.

It is equipped with a Case hydraulic lift with mounted two-furrow Case plough. The VA was also the first Case tractor to use the Eagle Hitch in 1949.

RIGHT: View from just below the driver's seat.

BELOW: Case was often referred to as the Rolls-Royce of tractors, and certainly they were beautifully finished.

Aircraft pulling

Taken in October 1942 by photographers working for the Office of War Information (OWI), the picture shows a B-25 bomber at North American Aviation (now part of Boeing) in southern California, being hauled along an outdoor assembly line by an International tractor. During World War II, tractors were regularly used for manoeuvring aircraft around military airports. They had the power and, of course, traction to be able to pull these giant planes.

The Fordson E27N Major

Based on the Fordson model N, and using the same engine and gearbox in a new casing, the E27N Major was introduced in 1945. During World War II, 94 per cent of all wheeled tractors made in Britain were Fordsons, and an advert of 1947 showed how keen the company was to keep sales up. In the attempt to help British farmers grapple with the problems of producing more food, Fordson introduced the Hydraulic Power Lift and an ever-expanding range of implements designed and built specially to operate with the Fordson Major tractor. Its slogan was: 'Go forward with Fordson'.

The model shown here also came equipped with a Roadless half-track system (pictured above right), which gave better traction, especially useful in muddy conditions.

Farmers found the Roadless half-track system a real help in the soft and muddy conditions common to the UK.

SPEC BOX	
Make/Model	Fordson Major E27N (kerosene)
Year	1945–1952
Engine	Ford 4-cylinder 267 ci (4.4 litres)
Horsepower	27
Power	Drawbar – 28.3 hp (tested) Belt – N/A

The very tall Fordson E27N ran on paraffin in England, but Perkins P4 and P6 diesel versions were also available.

The Farmall model H

After the initial success of the Farmall 'Regular' model, as it became known, International Harvester stepped up its quest to compete with the success that Ford was having with its tractors by introducing the larger model H to its 'letter series' in 1939. The Farmall H was ideal for farms up to 160 acres in size. It could pull two 14-inch bottom ploughs and proved especially useful for cultivating row-crop produce such as potatoes and sugar beet, cultivating up to 35 acres of row crops per day.

SPEC BOX

Make/Model	Farmall H (petrol)
Year	1939–1953
Engine	International Harvester 4-cylinder 152.1 ci (2.5 litres)
Horsepower	23
Power	Drawbar – 24.17 hp (tested) Belt – 26.20 hp (tested)

BELOW AND OPPOSITE: The Farmall H rose to the challenge of cultivating larger farms with great success.

The H was originally available with steel wheels and also rubber tyres, but with World War II about to envelop the world, rubber became scarce and so mainly steel was used until peace was restored. It was available in several different configurations: tricycle, adjustable wide-front or fixed wide-front, high-crop versions with optional hydraulics.

The H was produced up until 1953 and in its time became the top-selling tractor of all time in North America, even beating the Ford N8, which came a close second.

Chapter 4

THE WORLD'S WORKHORSE

After World War II, many European countries were devastated, swathes of buildings were bombed out, roads were destroyed and people demoralized. Germany in particular had been hit hard by British and American bombers, while Italy had taken a beating from the Germans, who had destroyed cities on their retreat from the Allies. Britain, of course, had been bombed heavily and many of its major cities were left in ruins.

As people gradually started to put their lives back into some kind of order, it was obvious that there was going to be a shortage of food – as if the rationing during the war had not been bad enough. Many of the pre-war tractor manufacturers had had to change their production lines during the war, to make guns, ammunition and other essential supplies for the war effort. But now that the war was over, it would be back to farm vehicle production once again.

As the 1950s approached, although tractors did not change too much in actual design, they did get bigger, more powerful, more comfortable, safer and better able to carry out a variety of tasks. And so the farmer found himself producing more without the expense of having to hire in gangs of people. It is interesting to note that the sum of money that the farmer had to spend on employing people fell by 35 per cent between the end of the war and the early 1960s. An interesting situation had occurred during the war. The farmers who had been left behind to work the land began to realize that what they really needed was machines that could carry out a multitude of tasks, rather than more people on the payroll. Now farmers found they had money available to spend on new machinery and they were getting a great deal of choice, which also included a variety of implements that would do the work of several people.

Fuel, of course, was an important factor too. Although LP (liquefied propane) gas was being used in the 1950s and 1960s, many farmers started adopting diesel as their main fuel, although diesel tractors didn't always start very well. In this case many turned to spark-ignition diesel engines, which were started on gasoline and could be switched to diesel once warmed up. By the 1970s, nearly all tractors used diesel.

As the decades passed, the tractor and the implements that could be attached to it became more and more sophisticated and easier to use.

And as the new millennium came closer, so the larger companies started to swallow up their rivals in time-honoured fashion. Worldwide corporations were created where several companies came under one umbrella. A lesson had been learned after they had battled through the recession of the late 1970s/early 1980s and then again in the early 1990s. Consolidation would allow them to fight these lean times better and give them more chance to survive any future downturn.

Cockshutt

The tractor and machinery company Cockshutt was based in Brantford, Ontario.

It was founded initially as the Brantford Plow Works by James G. Cockshutt in 1877, but then the name was changed to the Cockshutt Plow Company when it was incorporated in 1882. It was a family-run concern and they built up a reputation for quality products.

Production of the first Cockshutt tractor, the model 30, was delayed because of World War II, with the first tractor finally coming off the production line in 1946. The Cockshutt 30 became the first production-line tractor to be built in Canada.

It is interesting to note that, because Canada was part of the British Empire during wartime, Cockshutt's Brantford factory manufactured undercarriages for several types of British bomber, as well as building plywood fuselages and wings for the Anson training aircraft and the Mosquito bomber.

The 1946 model 30 shown here was powered by a gasoline, 4-cylinder, 153.1 ci (2.5 litre) Buda engine and had four forward gears and one reverse.

ABOVE and BELOW: Production of the Cockshutt 30 was postponed until after World War II, thus giving the company access to materials restricted during that period. Wanting to sell beyond Canada, the company set up agreements with companies in the USA: in one instance, the name was changed to the Co-op, and the livery took on a 'pumpkin orange' colour with black lettering.

SPEC BOX

Make/Model	Ferguson TE20
Year	1946–1948
Engine	Continental Z-120, 120 ci (2.0 litres)
Horsepower	23
Power	Drawbar – 20.70 hp (tested) Belt – 25.41 hp (tested)

The 'little grey Fergie' has made history on several occasions. Once a fleet of them helped to build levees to save the city of Wentworth from being flooded by the Darling and Murray rivers in Australia. A monument was erected to commemorate the event.

Ferguson TE20 1947

The Ferguson TE series of tractors was introduced in 1946 in England, although the 'Ferguson System' that the tractor was equipped with had taken considerably longer to develop. The designation TE stood for 'Tractor England' and the number should have denoted the horsepower, but in fact did not in this case.

In an attempt to get a serious volume manufactured, Harry Ferguson made a trip to America and demonstrated the tractor to Henry Ford, who agreed to manufacture a joint effort in Detroit under the name of Ford Ferguson. The badging would feature both names.

Ferguson wanted to manufacture the tractor in England and had agreed with Ford that work would begin at the Dagenham works in Essex. Unfortunately, Ford UK was reluctant, forcing Ferguson to look elsewhere. An agreement was finally struck with the Standard Motor Company in Coventry, who developed a new engine for the machine, which was also used in its Vanguard model automobiles.

When production finally started in 1946, it incorporated most of the new developments that Ferguson and his team had been working on. Many of the Coventry TE20s were shipped to America and Canada, under the designation TO20 (Tractor Overseas), although there was little difference in their design. Harry Ferguson joined with the Massey-Harris Company in 1953, the company adopting the name Massey-Harris-Ferguson, which would eventually become the Massey Ferguson that we know and love today.

Orsi Pietro & Figlio

Orsi Pietro & Figlio became one of the biggest producers of farm machinery of its time. The company's history goes back to 1881, to a small workshop in Tortona, Italy, where they started making small farm tools and ploughs. Soon the word spread regarding the excellent quality of their work.

It was Pietro's son Giuseppe who first had the dream of making farm machinery that would help the local farming community. Giuseppe joined his father at the turn of the century and by 1902 they had invented the first straw baler and by 1907 had constructed their first steam engine.

A gold medal at the prestigious International Exhibition of Richmond in England placed them on the international stage.

Following in his father's footsteps, Giuseppe's son Luigi then took the reins and it was he who instigated the production of tractors, competing now with the other well-established Italian companies of the period. Sadly things started to go wrong as Luigi died prematurely in 1936, leaving Giuseppe alone to run the company. The war intervened not long afterwards and, to make things worse, Giuseppe was kidnapped and held hostage for a large ransom. He died a few years after being released in 1948.

By the 1950s the company was struggling and even the attempt to find a partner proved hopeless. Finally, on 2 July 1964, the company which had once been one of the largest mechanical producers in the region was obliged to declare bankruptcy.

Shown on this page is a 1950 Argo, powered by a single-cylinder, horizontal, hot-bulb, semi-diesel 11,277 cc engine, with a power output of 60 hp at 670 rpm. It has six forward gears and one reverse.

LEFT: A rather tatty example of an Argo, but this is now a rare machine. Orsi was instrumental in pushing the production of hot bulb engines further than most companies. The Orsi name is not to be forgotten: where the factory once stood in the centre of the town of Tortona, there is now a museum dedicated to the company.

Landini 'hot head'

Giovanni Landini founded his company in 1884, in Italy, to make agricultural equipment. Prior to World War I, he also built internal combustion engines. Landini died in 1925 just as the company was preparing its first tractor prototype. His sons took over the business and saw the prototype through to completion. They followed up with their first complete tractor, which used a hot-bulb, diesel engine producing 30 hp. This became the precursor to the Landini 40 and 50 hp machines of the mid-1930s – the Velite, Bufalo and Super.

SPEC BOX	
Make/Model	Landini
Year	1950–1959
Engine	Horizontal single-cylinder, 4,312 cc, two-stroke, hot bulb
Horsepower	25–30
Power	Drawbar – 25 hp Belt – 30 hp

RIGHT: the 'hot bulb' engine of the Landini L25. Although this type of engine was not widely used, it has become a much-loved feature with tractor enthusiasts today.

Production stopped during World War II, but the L25 came on to the market in 1950, and also used a 4,312 cc, hot-bulb engine.

Sadly, these engines were now out of date and farmers began looking towards much more modern equipment. Although these machines remained in production until 1959, the company struggled financially and, in this same year, it was purchased by Massey Ferguson.

Consolidation of Ford

Built in England from 1952, the Fordson New Major E1A series of tractors would be the last to bear the Fordson name and badge. In 1961 Ford decided to consolidate the US and British companies and the two became one, under the same Ford badge and with a grey-and-blue colour scheme, although the Fordson name continued to be used in Britain until 1964.

Although the Major featured here is the diesel version, it could be adapted to run on most fuels. The tractor became a big success and went through several upgrades. The Power Major came in 1958, followed by the Super Major and the New Performance Super Major. All were made at the Ford Dagenham factory, although the Super Major was also sold in America as the Ford 5000.

SPEC BOX	
Make/Model	Fordson New Major E1A (diesel)
Year	1952–1958
Engine	Ford 4-cylinder 220 ci (3.6 litres)
Horsepower	40
Power	Drawbar – 34.2 hp (tested) Belt – 38.5 hp (tested)

RIGHT: The driver's view of the Fordson Major.

BELOW: The E1A New Major originally went into production with a Spanish-made diesel engine, although it was designed and built in Dagenham, England.

Deutz D15

Today's Deutz-Fahr company, now a part of the SAME Deutz-Fahr group, is closely associated with one particular name, Nicolaus August Otto. It was he who founded the Motorenfabrik N A Otto & Cie., which later became Klöckner-Humboldt-Deutz AG (KHD) in Cologne in 1864. Shortly afterwards, Otto invented the first 4-stroke combustion engine.

Manufactured from 1959 to 1964, the Deutz D15 was the smallest of the D series tractors. It was powered by a single-cylinder, 850 cc Deutz diesel engine and had a maximum speed of approximately 12 mph (19 kph).

The D15 Deutz tractor uses a Deutz diesel engine. These engines have a character of their own: they are simple, reliable, rugged and air-cooled. Being air-cooled means less maintenance as there are no liquids to change, no water pumps to fail and no seals to go wrong.

Fewer petrol versions of the Dexta were produced than diesels. The US version was the Diesel 2000, and other models included the industrial version, as well as a special design for the German market.

Fordson Dexta

Nearly ten years had passed since the end of World War II and it was obvious that the smaller tractor was now a requirement. Smaller farms needed a tractor that was suited to small-scale farming and the Ferguson 35 was a classic example of this. That's why the Fordson Dexta tractor was launched in 1957. This tractor was a completely new design, intended to compete directly with the Massey Ferguson 35. The Dexta was powered by a 30.5 hp, three-cylinder, direct-injection diesel engine, with an in-line fuel injection pump and pneumatic governor. There was a gasoline version, but it sold in much smaller numbers than the diesel alternative. In Germany a Dexta Special was made, a narrow model was also available and, in America, it was sold as the Diesel 2000.

The one shown here is, in fact, quite rare. It is an Industrial version, which is easily identified by the large mudguards that extend over the wheels. This tractor worked in a factory and often had to travel along a public road, and so these mudguards were a legal requirement.

Fordson branding was dropped after 1964 when all tractors reverted to the Ford name in the UK and the USA.

The 'little grey Fergie' becomes the new MF35

The Harris part of the Massey-Harris-Ferguson company name was removed at the end of 1957 and the legendary Massey Ferguson brand was born early in 1958. Initially the name was spelt with a hyphen, but that was removed later on. And so the little grey Ferguson FE35 borrowed its new red-and-grey colours from its new parent company and became the MF35 that same year.

The original 4-cylinder diesel version of the MF35 was replaced in 1959 with a 3-cylinder Perkins diesel engine. Massey Ferguson bought Perkins Engines of Peterborough, England, in 1959, Perkins having been the main supplier of diesel engines to the company for many years.

One advantage with the new engine was that it would start better in the mornings, a complaint many older farmers can recall even today. During its illustrious lifetime, the MF35 was also made in gasoline and gasoline/vaporising oil versions.

In January 1958 Sir Edmund Hillary became the first person to reach the South Pole overland since Scott and took with him three modified FE35s. He sent a telegram from Antarctica, which said: 'Despite quite unsuitable conditions of soft snow and high altitudes our Fergusons performed magnificently and it was their extreme reliability that made our trip to the Pole possible.' It doesn't get much better than that!

ABOVE: Without doubt a beautiful little tractor, the MF35 still showed off its heritage even though its mechanics had changed.

OPPOSITE: The early MF35s still had the Ferguson name badge on the front. Before long, though, the new MF badge replaced the one on the front, as here.

SPEC BOX	
Make/Model	Massey Ferguson MF35 (diesel)
Year	1960–1965
Engine	Perkins 3-cylinder 152.7 ci (2.5 litres)
Horsepower	37
Power	Drawbar – 32.13 hp (tested) PTO – 37.04 hp (tested)

Massey Ferguson MF135 – the tractor body gets a makeover

The early part of the 1960s saw another change and the smaller Massey Ferguson tractors – the 35, 50 and 65 – which by now were starting to look a little dated, were given a makeover.

The Red Giants, as they were commonly referred to, were now being produced for a world market and although the new 135 model used the same engines as the 35 – Continental gasoline or Perkins diesel – they inherited new styling with a radical new bodywork design.

Considered more reliable and more powerful than other tractors of the same period, they became very popular and sold well. The MF135 remained in production until 1976, although the orchard, vineyard and crawler versions were made until 1982.

LEFT: Two of the range pictured together: the MF165, the largest of the series (right), next to the smallest, the MF135.

BELOW: The MF135 was popular with small family-run farms. A real workhorse, it could manage everything from cultivators and manure spreaders to seeders and hay balers. Also available was a front-end loader, which added to the possibilities.

Although looking very different from the MF65, the MF165 was much the same, except for the new tinwork.

The 'Big Red Giant' improves with age

Over its lifespan, the MF165 improved in several important respects. The engine, for example, increased in power from 56.8 to 60 hp.

Although the Massey Ferguson MF165 was the replacement for the 65, besides its new looks it had changed little from its predecessor. During its 12-year lifespan, it did increase in power output and, in the early 1970s, dry element air cleaners, oil-cooled brakes, improved hydraulics and independent power take-off were all incorporated. Manufactured between 1964 and 1975, this mid-range Red Giant tractor could also be supplied with a Perkins 4-cylinder diesel engine, Perkins 4-cylinder gasoline engine or a 4-cylinder Continental gasoline unit, all with varying displacements. As with the smaller MF135, optional extras included such items as sprung suspension seat and a cigarette lighter.

The David Brown 950 Turbo Taskmaster Super

Built as an aircraft-towing vehicle, this 1960s David Brown tractor still has the looks of its 1940s parent. Originally sent to Farnborough to demonstrate its ability to tow a Victor bomber, which it did satisfactorily, it was commissioned into the RAF in 1962.

Forty of these David Brown 960 Taskmasters were built, of which only eight were the Super model, with double rear wheels. From 1963 to 1972, the tractor served at RAF Cosford and, from 1973 to 1997, she was with Marconi Defence Systems UK, before being purchased by a collector.

This tractor had a towing capacity of 44.6 tons; it was fitted with a 42 hp, 4-cylinder diesel engine and had a top speed of 19 mph (30 kph).

ABOVE: The business end of the machine shows its towing capacity. Note the double wheels and counter weights.

BELOW: Close-up of the enormously powerful 4-cylinder David Brown engine.

The Allis-Chalmers ED-40

The ED-40 would be the last Allis tractor to be made in England. Production started at the Essendine Factory in Rutland in 1960 and ended in 1968. Because sales of this model were slow, it was still possible to find a new ED-40 long after manufacture ceased. Of 4,000 units built, some 450 were exported to the USA.

The ED-40 was a heavier construction machine than its predecessor, the D272. It was fitted with a Standard Riccardo 2.3-litre engine, 8-speed gearbox, live hydraulics and the option of live PTO.

ABOVE: The ED-40's instrumentation was far more sophisticated than was usual in most tractors of the period.

BELOW: The ED-40 was the last of the line of tractors built by the UK subsidiary of the US Allis-Chalmers Corporation, after which the Essendine factory was closed down.

McCormick 414

The International Harvester Company of Great Britain Ltd was incorporated on 31 December 1906, with offices in London. But until 1939, International Harvester of Great Britain Ltd did not actually manufacture tractors, but only imported and assembled products from the USA and Canada. The first assembly plant in the UK was set up in Liverpool, near the docks at Orrell Park, in 1923. In 1938 a large works for full-scale manufacturing was built at Doncaster. After being requisitioned by the UK government during World War II, it was returned to IHC in 1946, when it started producing tractors, crawler tractors and farm implements.

The all-new McCormick International Harvester B-414 tractor was introduced at the Royal Show in 1961. It was available with IH 4-cylinder diesel or 4-cylinder gasoline engines, 'Vary-Touch' (the new IH fully 'live' hydraulic system with automatic draft control) and a large 10.5-gallon fuel tank. A one-piece removable radiator grille and radiator cap access panel was provided for easy service.

LEFT: Shown here is the diesel engine of the B-414, but there was also a petrol version.

BELOW: When the B-414 was introduced, the ads described it as 'the tractor with everything you've wanted most'.

The Fiat crawler

The 'BO' on the registration plate gives away the locality of this ploughing scene. Bologna in Italy has a big farming community and this crawler is ideal for the kind of heavy earth that is found there. The crawler itself is a Fiat 120C model, and although this image was taken in 2006, the crawler is from the late-1970s/early-1980s period – testament to the endurance of these powerful little machines, which can often be found working the fields all over Italy even today.

SPEC BOX	
Make/Model	Fiat 120C crawler
Year	1970–1980
Engine	Diesel
Horsepower	120 (88 Kw)
Power	N/A

Although it's small, the power of the Fiat 120C allows it to tackle the hardest of jobs. For example, the furrows in this field have to be dug deep. The crawler came with protection from the sun: a material roof was available and could be attached.

A Scandinavian story

The Finnish Valtra Company, today a part of the huge AGCO Corporation, has a long and winding history that goes back to 1832. In that year, Johan Theofron Munktell owned a mechanical workshop named Eskilstuna Mekaniska Werkstad, and it is here that the industrial history of both Valtra and Volvo Construction begins. The company supplied Sweden with its first steam railway engines and also started building traction engines. By 1905 Munktell had built its first diesel engine and in 1913 it built its first tractor. After World War I, and with exports to Russia ceasing, the company fell into financial difficulties. In 1932, its 100th anniversary, the company joined forces with Bolinder and became AB Bolinder-Munktell, or simply BM.

The two Bolinder brothers, Jean and Carl Gerhard, had started a mechanical workshop and foundry, J & C G Bolinder, in Kungsholmen, Stockholm in 1844. Initially the company manufactured steam engines, woodworking machinery and cast iron products. In 1893 it manufactured the first 4-stroke carburettor engine of its type in Sweden; semi-diesel engines were the company's biggest success, however. Carl's son Erik August took over the company at the turn of the century until his death in 1930.

AB Volvo was established in 1927 as a subsidiary of SKF, the

ABOVE: Munktell's first tractor, introduced in 1913, came with a power unit of up to 40 hp and weighed a massive eight tonnes.

BELOW: Powered by producer gas due to wartime shortages, the GBMV-1 went into production in 1943.

Swedish ball bearing company. In its first year it manufactured 297 cars. During World War II, Volvo manufactured tanks, which inspired its management to produce an agricultural tractor with a carburettor engine and a transmission sourced from Bolinder-Munktell. Tractor production continued after the war and in 1950 AB Volvo bought out Bolinder-Munktell. Tractor production was transferred to Eskilstuna in order to free up space for expanded car production. Throughout the 1950s the plant produced both green Bolinder-Munktell tractors and red Volvo tractors. The now famous 'hot-bulb' engines entered the history books in 1952, with the introduction of the Bolinder Diesel Series, and in 1958 the official company colour became red, with the brand name officially changed to Volvo BM in 1973. During the 1970s Volvo wanted to concentrate on excavators and initiated talks with International Harvester. This didn't work out and so it turned instead to Finnish company Valmet.

The Volvo 470 Bison, introduced in 1961, had a reliable and strong Bolinder engine and was ideal for large farms.

Valmet, or Valtion Kivääritehdas as it was known, had started out as a government factory producing rifles in Tourula, Finland. During World War II it was kept busy, but after the war, under the treaty agreement, it was no longer allowed to produce armaments. The company, therefore, moved into other fields, such as the production of metal tools, dentist's chairs and the like, and became part of the State Metal Works (Valtion Metallitehtaat, VMT).

By 1951, VMT, now known simply as Valmet, started work on a tractor project. Tools were modified at the Tourula factory and by 1952 a test series of 75 machines had come off the production line. With the success of the Valmet 33 in the mid-1950s, the company went from strength to strength and was soon exporting to other countries, in particular Brazil, where even today the company has a huge presence.

The Valtra name had been registered in 1963 and in 1970 it was used to brand a range of implements for the Valmet tractors. The company began to expand rapidly. When Volvo decided to

cease tractor production, Valmet, through a company called Scantrac, bought that part of the company out, although components would still be supplied by that source. A new tractor designated Volvo BM Valmet was introduced in 1979 and as the years progressed more innovative machines were introduced. Sadly the recession of 1991/1992 hit the company hard and a restructuring exercise had to be instigated. Sisu acquired the tractor, forest machinery and material handling divisions in 1994. Brand name changes were now in discussion and with the company having been split - the Valmet name was used by the paper production works - the tractors were given the Valtra name. In the following years Sisu was acquired by Partek, a chalk mining company, and the name Valtra Valmet was launched. Finally in 2001, the company became simply Valtra, although that was not the end of our story. Partek was taken over by Kone Corporation in 2002, which subsequently sold off the tractor and forestry machinery to the AGCO group.

In 1964 Valmet introduced the 52 hp model 565 with its synchronized transmission.

The Versatile 276 is a 'bidirectional' tractor, so called because the operator's console can turn 180 degrees, allowing the tractor operator to work in either direction.

Versatile 276

Versatile Manufacturing Ltd was incorporated in 1963, and just a year later a new factory was built in Fort Garry, Winnipeg. In 1966 it entered the four-wheel-drive tractor market with the D100, a 100 hp tractor fitted with a Ford 6-cylinder diesel engine. Other models were to follow, most of which were large and powerful. In the 1970s Versatile was an independent operation that had 70 per cent of the four-wheel-drive market.

An innovative tractor of the 'push-pull' design was introduced in 1977, called the Model 150. This was replaced in 1984 by the Model 256, which churned out 85 hp, and an even more powerful 100 hp model, the 276 (shown on this page) followed in 1985.

The company was sold to Cornat Industries of Vancouver in 1977, and the world market was to open up in 1979 when a deal with Fiat was struck to supply tractors to the European market. Several models of differing horsepower were sold under the Fiat badge, the 44-28 being one. These were all painted in the sandy-orange Fiat colour.

Lamborghini

Ferruccio Lamborghini, although better known for his amazing sports cars, started building tractors in Italy in the late 1940s. Initially, the tractors were built using a mixture of surplus military hardware left over from World War II. By 1954, Lamborghini was building his own engines and this was when he also started building his high-performance sports cars. In the late 1960s, Lamborghini lost his passion for tractors and the company was formally acquired by SAME in 1971. Today the Lamborghini name is part of the SAME Deutz-Fahr Group and produces a wide variety of tractors.

Seen here is the Crono 554-50, a model introduced in the 1990s. It was powered by a 3-cylinder, 183.1 ci (3-litre) diesel engine, that produced 50 hp.

The series of Lamborghini tractors is extremely popular in its home country of Italy. The Crono is a mid-range tractor that performs well for a variety of tasks on medium-sized farms.

Chapter 5

TRACTORS IN THE 21ST CENTURY

By the time the new millennium dawned, many tractor manufacturers had combined forces to prepare themselves for the challenges ahead. Even large companies like Massey Ferguson had been taken under the wing of a huge conglomerate known as AGCO, which also owned Challenger, Fendt and Valtra. The SAME Deutz-Fahr Group also incorporated companies such as Lamborghini and Hürlimann, while Fiat and Ford tractors had become CNH (Case New Holland). Many of the tractors from the different companies within a group would use the same mechanics, thereby cutting down on manufacturing diversity and excessive labour costs.

By now, this kind of restructuring was not just seen in agriculture but in other businesses too. Yet in spite of the constant improvements in efficiency that the tractor had brought to agriculture over the previous century, it had not solved the world's food problems.

If anything they had become worse. In the West the problem was not a lack of food but an excess, with obesity affecting large chunks of the population and on the increase. Meanwhile, famine continued to threaten the lives of people in the Third World. Could the tractor, having revolutionized the production of food in the developed world, now help to establish a more even distribution of health and nourishment throughout the planet?

Tractors were now cleverer than ever and they enabled just one operator to carry out a multitude of chores. Every moment of the farmer's life seemed to be calculated by a computer. Crops were cut and harvested according to the weather forecast that the computer gave. The way the fields were cut was calculated by a computer and the machine that did the cutting was controlled not by the driver but by a computer. Hi-tech instrumentation was now the order of the day.

AGCO

Although it is a relatively new concern, AGCO can boast a brand heritage that goes back to the mid-1800s. The AGCO group was established in 1990 with the acquisition of the Deutz Allis Corporation from the German-based Kloeckner-Humboldt-Deutz AG, which in turn had purchased portions of the Allis-Chalmers agricultural equipment business five years earlier. Since that time, AGCO has become a farm machinery company with fingers in a great many agricultural pies around the world.

It boasts such names as Massey Ferguson, Challenger, Fendt, Valtra and, of course, AGCO tractors as part of the group. Seen here is the AGCO DT275, top of the range in the DT series. It is powered by a 514 ci, 6-cylinder, AGCO Sisu Power engine with Electronic Engine Management 3 (EEM3).

ABOVE: The cab of the DT275, which is one of the quietest of any tractor, allows excellent all-round visibility, while also being comfortable enough for a day's work.

BELOW: The DT275, with its twin front and rear wheels, is seen teaming up with a combine harvester.

The Challenger

Caterpillar Inc. created the Challenger tractor, the world's first rubber-tracked agricultural tractor, in 1987. The original model was a Challenger 65 featuring the Mobile-Trac System (MTS), which consisted of rubber tracks and a suspension system. In 1995 Caterpillar introduced the first and smaller row-crop, tracked machines, the Challenger 35, 45 and 55, which were designed to handle many of the tasks that the larger machine was too big for.

The 2011 MT800C series of rubber-tracked special application tractors are designed for heavy groundwork operations. The MT875B broke the world record for the most land tilled in 24 hours with a custom-made, 46 ft (14 m) disc harrow; it tilled 1,590 acres (643.5 hectares). The tractor consumed 180 l/km² of diesel fuel. Featured below are three different Challenger models: the massive-tyred MT965C is on the right, the smaller MT765C at the top and the MT800C on the left. In 2002, Caterpillar sold its tracked vehicle line to AGCO.

ABOVE and LEFT: Taking their lead from the MT800C series tracked tractor (above), the MT900C tyred machines (left) are among the most powerful articulated tractors on the market.

Massey Ferguson

There are currently more Massey Ferguson tractors than any other make in the world. The brand's history dates back to 1847 when Daniel Massey opened a workshop to build simple farm implements in Newcastle, Ontario. As outlined earlier, Massey and Harris were leading names in farming equipment when they combined assets in 1891 to create the Massey-Harris Company. This rapidly grew to be the biggest agricultural equipment company in the British Empire.

In the 1930s, Massey-Harris brought out the first self-propelled combine harvester as well as a pioneering four-wheel-drive tractor. Innovation was always key in the market and, for decades, it remained locked in competition with Ford to produce the ideal farm tractor, one which would help provide improved yields at lower costs, in other words cheaper food.

In 1953, Massey-Harris joined forces with a brilliant Irish engineer named Harry Ferguson, who had revolutionized tractor design with his three-point hitch. For the first time ever, tractor and implement could work as one – a concept that still applies today on virtually all agricultural tractors.

The new association was called Massey-Harris-Ferguson Limited, but five years later the name was shortened to Massey Ferguson, creating one of the world's major forces in farm equipment. The company swallowed up other businesses such as Sunshine of Australia (1955), Italy's Landini (1959) and Perkins Engines of England (1959), and, for a brief few years, even made snowmobiles. But in 1995, Massey Ferguson was bought up by the American AGCO Corporation, which continues to turn out shiny new models such as the ones you see on these pages.

RIGHT: The Massey Ferguson MF8600 Series is their biggest, most powerful tractor range to date. These machines not only have a seriously modern look, but also combine a powerful 6-cylinder AGCO SISU POWER engine and Dyna-VT transmission.

BELOW: The MF3400 series machines are small and compact but will tackle a multitude of tasks around the farm. They come with or without a roof.

The Fendt brothers

The Fendt brothers, under the guidance of their father Johann Georg, started building tractors in a blacksmith's shop in Germany between the wars (see page 47).

The company's slogan is 'For a head start, drive a Fendt' and it has a number of firsts to its name: the first 'truly affordable' European tractor (1930); the first tractor with an upright radiator (1938); the first tractor with up to 150 hp (1976); and the first tractor to be capable of 40 kph (25 mph – 1980) and 50 kph (32 mph –1993). In 1961, the 100,000th Fendt tractor rolled off the production line, the Farmer 2, and there have been many more since.

BELOW: The legendary 965 Vario was launched in 1995 and today, when farmers are confronted by ever-growing field sizes, the 900 Vario range offers a huge appetite for work plus many time-saving devices.

After the Second World War, the company enjoyed rapid expansion under the management of the Fendt brothers, Hermann, Xaver and Paul. The Dieselross line continued and in 1958 the new Favorit 1 was put on the market. The new Farmer range was soon to be added to the catalogue, followed by the ingenious GTA System Tractor in 1984. The trendsetting system vehicle Xylon (110 to 140 hp) was introduced to the market in 1995 and just two years later the company became part of the AGCO Corporation. Today, Fendt has an incredible range of modern tractors, which can accommodate most duties on and around the farm.

ABOVE: The mid-range Fendt 818 Vario has become the best-selling, compact, high-horsepower tractor in Europe.

BELOW: Released in 1998, the Fendt 700 Vario is now in its third generation. It has been upgraded and had its range expanded. This all-rounder now has a capacity ranging between 130 and 180 hp.

SPEC BOX

Make/Model	Fendt 936
Year	2011
Engine	Deutz 6-cylinder, common rail, turbocharged, intercooled
Horsepower	330
PTO	Rear: electro-hydraulic Speed – 540E/1,000

Valtra

A brief account of the intriguing story of the Valmet tractor company has already been given (pp.86–87), and as mentioned, the roots of Volvo BM and its predecessors Bolinder and Munktell lie deep in Swedish industrial history.

Today, Valtra tractors are built at Suolahti, Finland, and Mogi das Cruzes, Brazil, in the most advanced factories in the industry. With sales outlets that cover over 75 countries, Valtra is now a worldwide brand of the AGCO Corporation, which in turn is the third-largest manufacturer of agricultural equipment in the world.

Valtra tractors are made to order and custom built to the requirements of their customers. In consultation with the authorised local dealer, the customer can request what features he wants his tractor to have. The order is then forwarded to the factory, where the machine is constructed in accordance with the wishes of the customer, then delivered.

ABOVE: The T series Valtra tractors are suited to farmers, contractors and forestry specialists. Flexibility is the key with this multi-purpose control panel.

LEFT: The Valtra N series is versatile and agile, which it has to be to tackle all the tasks it may be called upon to perform.

SPEC BOX	
Make/Model	Valtra T 131/161
Year	2011
Engine	Sisu 6-cylinder, common-rail
Horsepower	148/170
PTO	Type – Rear: electro-hydraulic Speed – 540/540E/1000 /1000 HD

The legacy of John Deere

The story of John Deere, who became a blacksmith and inventor, follows closely the settlement and development of the Midwestern USA. In the 19th century homesteaders looking for land to own and farm considered this the golden land of promise.

When he fashioned a polished-steel plough in his Grand Detour blacksmith shop back in 1837, John Deere gave the pioneer farmers the ability to cut good, clean furrows through sticky Midwest prairie soil. His move to Moline, Illinois in 1848 was smart, as operating from there he was able to take advantage of improved water supply and transportation links. In the following years as the business grew, the company was able to expand both its repertoire of implements and its sphere of operations.

The American Civil War saw agricultural production increase to help feed the armies on both sides, and in 1863 the company introduced the Hawkeye Riding Cultivator, the first Deere implement adapted for riding. Even the Panic of 1873 and the grasshopper infestation that followed couldn't stop the company from growing. In 1875 the Gilpin Sulky Plow gave farmers a seat to sit on and became the company's best-selling product of the 19th century.

Deere died in 1886, but his legacy would live on as family members took over the mantle on his behalf. In 1918, at the end of World War I, Deere bought out the maker of the Waterloo Boy tractors. But the golden years of agriculture suddenly came grinding to a halt as workers went on strike and massive unrest spread throughout the farming industry.

ABOVE: The new 7R series of John Deere tractors comprises five new models which share two new engines.

LEFT: Putting its front loader to good use, the John Deere 6R.

OPPOSITE: The massive 9630 model, which features a 530 hp, 13.5-litre engine and 18-speed automatic powershift transmission.

LEFT: The view from the cab of the John Deere 5GV, showing how easily it copes with narrow row spacings.

BELOW: Showing just how useful the optional integrated front hitch can be, this John Deere 6210R is a multi-tasking marvel.

OPPOSITE: The huge John Deere 9630T tracked tractor, with its six-cylinder, 824 ci (13.5-litre) engine.

In 1923, Deere introduced the model D, the first 2-cylinder, Waterloo-built tractor to bear the John Deere name. It became a great success from day one and remained in production for 30 years. This was the beginning of the tractor company we know today. Today, John Deere is a major force in the agricultural world, with a line-up of machinery that covers every need of the modern farmer, from lawnmowers to combine harvesters, balers and more. John Deere tractors range from the compact and nimble 5GV to the gargantuan, 530 hp 9630T tracked tractor (far right) – with all sizes in between.

SPEC BOX

Make/Model	John Deere 9630T (track)
Year	2011
Engine	John Deere PowerTech™ Plus, 824 ci (13.5 litres)
Horsepower	530
PTO	Type – Rear: electro-hydraulic Speed – 1000

SAME Deutz-Fahr

More than 80 years have passed since Francesco and Eugenio Cassani submitted their first project for a tractor to be powered by a diesel engine. The hard work of these two individuals and their exceptional technological intuition, supported by knowledge of what farmers looked for in a tractor, led to the setting-up of a company whose quality products won prestige on the international stage. Today, the SAME Deutz-Fahr group is one of the biggest manufacturers of agricultural machinery in the world. Under its umbrella, there are some very famous tractor names and a history to rival anyone: Deutz-Fahr, Lamborghini, SAME and Hürlimann.

ABOVE: The Agrotron X model is a versatile modern workhorse.

OPPOSITE: The Agrotron X uses a Deutz, four-valve, turbo-diesel engine, which is both efficient and economical.

In the 1990s, with recession biting, standardization was on everybody's lips, but Francesco Carrozza, president of SDF, saw it differently. 'The fabric of the suit should be the same, but the cut and the tie should be different,' he said. More than a decade later, SAME Deutz-Fahr tractors continue to have a style and personality all of their own

Having sold over 50,000 units in a number of versions and with varying equipment options, the Agrotron series is one of the most popular tractors in Europe.

SPEC BOX

Make/Model	Deutz-Fahr Agrotron X 720
Year	2011
Engine	Deutz 6-cylinder, turbo-diesel with common-rail technology
Horsepower	270
PTO	Type – Rear: electro-hydraulic Speed – 540 E / 1000

The SAME Dorado

SAME machines have a particular character that caters to the Italian market. For example, the Dorado line is aimed at the specialist farmer and comes in 2- and 4-wheel drive versions. A compact but versatile tractor, ideal for farmers working in orchards and vineyards, it offers configurations that can be tailored to individual needs.

ABOVE: The compact Dorado F-S 70-100.

BELOW: The latest generation of SAME tractors, like the Iron here, use an advanced stepless variable transmission system.

The SAME Iron

Not top of the range but still a heavyweight within the group is the Iron model, 'a model designed for all needs'. With versions ranging from 130 hp to 220 hp, and a variety of implements to go with them, the Iron models have the versatility and power to take on most jobs around the farm.

The SAME Iron 130 and 185 are the latest generation of medium/high power tractors with which SAME brings design technology into the future, in an evolutionary process in which the tractor is transformed from a simple agricultural vehicle into a versatile, productive and extremely comfortable workstation.

The SAME Krypton crawler

The SAME Krypton crawler comes in two models, the 70 and 80 hp, specifically designed to meet the needs of specialist farming on sloping and difficult terrain.

These are extremely compact and highly manoeuvrable machines, which are ideal for the tight confines of vineyards, greenhouses and other speciality crops that require precision cultivation.

They perform a wide range of operations, such as crop spraying, shredding and mulching, row and inter-row cultivation and pruning. The safety frame can be reclined to reduce the height of the vehicle for working under canopies and foliage.

A SAME Krypton crawler at work in a vineyard.

The slimline Hürlimann XS is designed specially for fruit and wine producers.

Hürlimann

Hürlimann has always prided itself on producing quality rather than quantity. Today it is part of the SDF group and offers a wide range of machines. The powerful little Prince model is a modern interpretation of that first Hürlimann tractor. Capable of field work and obviously most at home around the farm, the Prince is reliable and pretty agile.

The little XS, with its narrow track and manoeuvrability, is designed for the fruit and wine farmer, while in the mid to large tractor category, the XM range offers versatility with the power of the Deutz 4- or 6-cylinder engines.

At the large end of the scale, the XL range is the flagship model, a tractor that can put in long hours with an air-conditioned cab. Power ranges from 130 hp to 180 hp.

The XM is a mid-range tractor, which can cope with most farm chores.

SPEC BOX

Make/Model	Hürlimann XM 100; 110; 120
Year	2011
Engine	Deutz 4-cylinder 4,038 cc or 6-cylinder 6,057 cc
Horsepower	112–126
PTO	Type – Rear: electro-hydraulically operated with modulated pushbutton Speed – 4-speed: 540; 540 Eco; 1000; 1000 Eco

The Hürlimann Prince is a multi-purpose machine.

The Hürlimann XL has a Deutz 6-cylinder, turbocharged, intercooler engine.

Lamborghini R1

The Lamborghini R1 might be small and compact, but it has the character of a big tractor. Versatile and multi-purpose, the R1 can handle open field work just as well as small farm, hobby farming and specialized crop applications.

Reliable, quick and agile, the R1 is well suited to greenhouse, plant nursery and park maintenance work. Low-platform versions are also available for orchard and vineyard work and for close-to-the-ground crop applications.

ABOVE: The Lamborghini R1 provides good performance on difficult terrain.

BELOW: Equipped with a specialist fuel injection system and fuel feed circuit, the R6 Deutz engine can run happily on the latest biofuels.

Lamborghini R6

The Lamborghini R6 is robust, reliable and technologically advanced. It has an innovative engine management system that a continuous analysis on information received from the engine, sensors and cab controls, and responds by delivering the power needed effectively at any given moment. Not only does this make the R6 more effective in its various tasks, but fuel consumption is significantly reduced too.

Lamborghini R8

Like its cars, Lamborghini tractors look like they mean business. The flagship R8 range has the power to back up its stylish exterior, but inside the cab all the grunt is cushioned by supreme comfort, including an operator seat that has an air suspension system that adapts automatically to the weight of the occupant.

SPEC BOX

Make/Model	Lamborghini R8.270
Year	2011
Engine	Deutz common rail, 6-cylinder, turbo intercooled
Horsepower	275
PTO	Type – Rear: electro-hydraulically operated, push-button control Speed – 540 Econ and 1000

The R8 is designed to allow full control with just the lightest of touches even with ploughing.

Landini

Landini is the standard-bearer of the 'Made in Italy' concept. Landini was established back in 1884 and the brand has evolved over the years. In 2008 Landini launched the Power Mondial range, featuring two different transmission models, Techno and Top. The Techno offers a 15-speed mechanical transmission, which can be combined with a creeper and mechanical reverse shuttle to achieve 20 forward and 20 reverse speeds. The Top version features a 12-speed transmission combined with a hydraulic three-shift option, which triples the gears of the basic transmission under load to 36 forward and 12 reverse. A creep transmission and a hydraulic power shuttle extend flexibility further still to 48 forward and 16 reverse speeds.

SPEC BOX

Make/Model	Landini Power Mondial 120
Year	2011
Engine	Perkins 4-cylinder, turbo-aftercooler
Horsepower	112
PTO	Type: Rear: electro-hydraulic Speed – 540/750 or 540/1000

The Landini Power Master 230, first produced in 2005, one year after the company's 120th anniversary.

ABOVE: The Landini Power Mondial is designed for a variety of tasks.

LEFT: The controls, to the right of the seat, are easily accessible.

Seen here at work is the top-of-the-range TTX tractor.

McCormick

Now part of the Argo group, McCormick continues to produce a comprehensive range of tractors, centred on machines that bring a high level of technology to open field work. The MC series features the latest in engine efficiency and operator comfort, exemplified by the powerful, 126 hp MC130 with 16x12 transmission.

SPEC BOX	
Make/Model	McCormick MC130
Year	2011
Engine	Perkins 4-cylinder 1104D-44T
Horsepower	126
PTO	Type – Rear: Auto PTO control Speed – 540/1000

The McCormick MC series has had a makeover, including a restyled four-post cab.

Under the bonnet of the MC series is the new Tier3-compliant, 1104-D turbo-after-cooler Perkins engine, which delivers up to 126 hp.

The versatile MC100.

CNH

The Case New Holland tractor story is a little more complicated than it seems and involves Ford, Fiat and several other companies too. So to keep things up to date, here are the most recent organizational arrangements: CNH Global N V is a full line company which has operations in both the agricultural and construction equipment industries. CNH's operations are organizedinto three business segments: agricultural equipment, construction equipment and financial services. The company was created on 12 November 1999 through the merger of New Holland N V and Case Corporation. CNH is majority-owned by Fiat Industrial SpA.

The Gold Signature Edition Steiger tractor, made in a limited run of just 50 units, was built by Case IH to commemorate the 50th anniversary of Steiger tractors in 2008.

One of the new generation Magnum tractors.

Magnum family

The Case IH Magnum range was developed in response to demands from farmers for more power and productivity, to handle their ever-increasing workload at greater speed and, importantly, with better fuel efficiency.

The Magnums feature Case IH Selective Catalytic Reduction (SCR) technology, which is designed to give the best possible performance for the lowest possible operating cost.

Power and productivity have been increased, along with a new driver environment that offers a level of comfort that could only have been dreamed of even a decade ago.

'Float through your field to maximise crop yields,' says the advertising. The Case IH Steiger Quadtrac features four individually driven oscillating tracks and other innovative designs that undeniably provide a smooth ride. Optimal ground pressure, superior flotation and better traction – all this and reduced compaction too. Steiger tractors have always featured a longer wheel base with built-in weight, giving better traction and more front-end weight.

ABOVE: The Farmall name is deeply rooted in Case IH brand history. Its tractors have been around since 1923. This is the 2011 Farmall C.

LEFT: A mean monster of a machine, the Case IH Quadtrac 600 holds the 24-hour ploughing world record.

The hydrogen tractor

Experts have long agreed that future driver systems for tractors and combines must be based on electricity, so New Holland presented its drive concept for the future in the form of its NH2 tractor. This hydrogen-powered tractor made its German debut at Agrotechnica 2009.

Unlike cars and industrial trucks, batteries are not a suitable power source for this type of vehicle because of the amounts of power that they use. The concept of the energy-independent farm, presented by New Holland, aims to make farms independent of external energy suppliers. The electricity which powers the NH2 is produced from hydrogen using fuel cells. The NH2 operates almost silently and emits only water and steam.

ABOVE: The cab of the hydrogen tractor offers a degree of comfort compatible with a modern automobile.

BELOW: The first steps to self-sufficiency. This tractor is fuelled by hydrogen, made and stored by the farmer who owns it, using renewable sources.

New Holland T9 series

2011 saw New Holland completely redesign and upgrade its American-built T9 tractor range. The T9 670 (on this page) is the brand's most powerful tractor, and is available in two versions: a standard and a heavy-duty frame to satisfy all farming needs. The standard 36-inch frame has been developed with row-crop producers in mind and it is equally at home in a mega-dairy situation or on farms where extensive road transport is undertaken. The heavy-duty frame model is the most powerful 4-wheel-drive tractor on the market and is perfect for extensive small grain growers, owner operators and large arable farmers.

CLAAS

It was 1913 when August Claas first started his business in Clarholz, Westphalia, Germany, and by 1919 he was producing straw binders at his new premises in Harsewinkel. The company began the development of its first combine harvester, which was specifically suited to European conditions, in 1930. It is this product, the combine harvester, that made the name Claas so famous in agricultural circles around the world.

The company acquired a majority stake in Renault Agriculture in 2003 and then acquired the remaining part in 2008, which expanded their product range into tractors. Today it has a formidable tractor range, which includes the Xerion and Arion series of tractors shown here. The Xerion is a large, powerful 4x4 tractor using a 3800 or 3300 CAT C9 six-cylinder turbocharged engine, while the smaller Arion boasts comfort for the driver while delivering up to 175 hp, if required, from the 640 top-of-the-range model.

ABOVE: Armed and dangerous, this Claas Arion tractor is powered by a highly developed engine and advanced transmission technology.

RIGHT: The Xerion 3800.

JCB

JCB are the initial letters of the names of the founder of the company – Joseph Cyril Bamford. He started the company in Uttoxeter, England in 1945, in a lock-up garage, equipped with a 50-shilling (roughly equivalent to $4 or €3) welder. His first product was a trailer made from war surplus materials.

The turning point for JCB came when he developed the first backhoe loader in 1953. Years before his time, Bamford wanted his new factory to be a place where the workforce and nature could thrive together, and so he created a wildlife sanctuary on the site, which has become an award-winning home to a diverse range of flora and fauna, including fish, birds, insects, hardwood trees and even orchids. Proof indeed that man, machine and nature can work in harmony.

SPEC BOX	
Make/Model	JCB Fastrac 8000 series
Year	2011
Engine	Sisu 6-cylinder, 513.76 ci (8.4 litres)
Horsepower	256–280
PTO	Type – Rear: Fully independent, electro-hydraulic Speed – 540/1000

INSET: The JCB 527 Agri Loadall comes with a variety of agricultural attachments.

BELOW: The JCB Fastrac's suspension gives a steady ride, however bumpy the terrain.

The Fastrac is fitted with a Stage IIIB (Tier 4i) Sisu engine, which supplies good power and torque.

Although best known for its excavators, which are extensively used in the construction industry, JCB has always been influential in agriculture. In 1990 it introduced the Fastrac tractor, one of the first tractors capable of travelling at speed on the road. The Fastrac has since evolved into the ultra-modern 8310, pictured here.

Anthony Bamford took over as JCB chairman from his father in 1976. The company went from strength to strength with record sales and a 17 per cent share of the total world market in 1984. In 1990 Anthony Bamford received a knighthood from the Queen.

Zetor

The Zetor Tractor Factory Company, established in 1946 in Brno, Czech Republic, has been producing quality-built, top-value tractors for the agricultural industry around the world, and has become one of the largest tractor manufacturers in Europe. The continuing success and growth of the company is based on its traditional research and development resources. Zetor designs and produces most of its tractor components, including the engines, which are among the most fuel-efficient in the industry. Down through the years the keynote of Zetor's success has been tough, reliable tractors at an affordable price.

In 1960, it produced one of the first hitch-hydraulic systems (Zetormatic), with full-position, draft and mixed control capability. Later it became the first tractor company to manufacture a fully integrated safety cab with insulated, rubber-mounted suspension – an idea that would be adopted by all other manufacturers. Again on current models, (even for the smallest 43 hp model), Zetor pioneered a flat, vibration-insulated, rubber-mounted platform (operator's station).

The Zetor tractor has been a big success. The new range, including this Proxima 95 Power, have all been updated and upgraded.

Zetor models for 2011 were further refined, while still including the Zetor easy-to-maintain chassis. Engines were more efficient and the transmission gave easier gear selection, while offering heavier-duty front and rear axles. Zetor tractors continue to be rugged and tough machines that play on their exceptional value-for-money theme.

BELOW: The Forterra series includes the 115 hp model.   RIGHT: The controls of the Proxima 95.

Kubota is well known for its small tractors, but in fact it also produces larger machines such as this M series 9540.

Kubota

The Kubota Corporation of Osaka, Japan was established in 1890 and has become an international brand leader. The company has subsidiaries and affiliates that manufacture and market their products in 130 different countries around the world.

The Kubota tractor has its roots on the Japanese farm and although they are smaller than the ones in America, the requirement for high performance and powerful manoeuvrability is the same. Kubota introduced its first tractor in 1969, filling a market black hole in the USA; the Kubota 21 hp L200 became an overnight success. As a result Kubota Tractor Corporation was formed in 1972, with offices in Torrance, California. The company introduced its first 12 hp, 4-wheel-drive machine in 1974. Kubota Manufacturing of America was established in 1988 in Gainesville, Georgia as Kubota's North American manufacturing base.

SPEC BOX

Make/Model	Kubota M130X
Year	2011
Engine	Kubota, 4-cylinder/turbo with intercooler
Horsepower	139.9 (Gross)
PTO	Type – Rear: Live-independent, electro-hydraulic Speed – 540/1000

Kubota is one of the world's largest manufacturers of diesel engines under 100 hp and is recognized as a world leader in compact tractors and diesel ride-on mowers. In 2010 the company celebrated 120 years of business and a new Kubota group slogan was created: 'For Earth, For Life'.

Another powerful agricultural machine from Kubota is the M130X model with its high-performance hydraulics giving a 5,800 kg lift capacity, and independent hydraulic PTO.

Index